Henry Jordaan and Manfred Max Bergman (Eds.)

Toward a Sustainable Agriculture: Farming Practices and Water Use

Series on Frontiers in Sustainability: Volume 1

selected papers from
5th World Sustainability Forum 7–9 September 2015, Basel, Switzerland

MDPI

Series Editor
Manfred Max Bergman
Department of Social Sciences
University of Basel
Switzerland

Guest Editors

Henry Jordaan   Manfred Max Bergman
Department of Agricultural Economics   Department of Social Sciences
University of the Free State   University of Basel
South Africa   Switzerland

Editorial Office
MDPI AG
St. Alban-Anlage 66
Basel, Switzerland

Publisher
Shu-Kun Lin

Managing Editor
Zinette Bergman

Production Editor
Seline Reinhardt

First Edition 2017

MDPI • Basel • Beijing • Wuhan • Barcelona • Belgrade

Vol. 1 ISBN 978-3-03842-330-0 (Hbk)   Series ISBN 978-3-03842-332-4 (Hbk)
Vol. 1 ISBN 978-3-03842-331-7 (PDF)   Series ISBN 978-3-03842-333-1 (PDF)

Table of Contents

List of Contributors

Reza Arjomandi Faculty of Environment and Energy, Science and Research Branch, Islamic Azad University, Tehran 1477893855, Iran.

Yonas Tesfamariam Bahta Department of Agricultural Economics, University of the Free State, Bloemfontein 9301, South Africa.

André Daccache International Centre for Advanced Mediterranean Agronomic Studies (CIHEAM-IAM.B), Valenzano 70010, Italy.

Najmeh Daryaei Department of Agricultural Development, Science and Research Branch, Islamic Azad University, Tehran 1477893855, Iran.

Daniel El Chami Timac Agro Italia, Ripalta Arpina 26010, Italy.

Antonie Geyer Department of Agricultural Economics, University of the Free State, Bloemfontein 9301, South Africa.

Jamal F. Hosseini Department of Agricultural Extension and Education, Science and Research Branch, Islamic Azad University, Tehran 1477893855, Iran.

Andries Johannes Jordaan Department of Agricultural Economics, University of the Free State, Bloemfontein 9301, South Africa.

Henry Jordaan Department of Agricultural Economics, University of the Free State, Bloemfontein 9301, South Africa.

Willem Lombard Department of Agricultural Economics, University of the Free State, Bloemfontein 9301, South Africa.

Frikkie Alberts Maré Department of Agricultural Economics, University of the Free State, Bloemfontein 9301, South Africa.

Hendrik Petrus Maré Centre for Sustainable Agriculture, University of the Free State, Bloemfontein 9301, South Africa.

Nomalanga Mary Mdungela Department of Agricultural Economics, University of the Free State, Bloemfontein 9301, South Africa.

Mehdi Mirdamadi Department of Agricultural Development, Science and Research Branch, Islamic Azad University, Tehran 1477893855, Iran.

Victoriano Joseph Pascual Department of Tropical Agriculture and International Cooperation, National Pingtung University of Science and Technology, Pingtung 91201, Taiwan.

Sejabaledi Agnes Rankoana University of Limpopo, Department of Sociology and Anthropology, University of Limpopo, Sovenga 0727, South Africa.

Samad Rahimi Soureh Agricultural Planning, Economics, and Rural Development Research Institute, Tehran 1598637313, Iran.

Walter van Niekerk Department of Agricultural Economics, University of the Free State, Bloemfontein 9301, South Africa.

Yu-Min Wang Department of Civil Engineering, National Pingtung University of Science and Technology, Pingtung 91201, Taiwan.

About the Editors

Dr. Henry Jordaan received his PhD in Agricultural Economics from the University of the Free State, Bloemfontein, South Africa. He is currently a Senior Lecturer in the Department of Agricultural Economics at the University of the Free State. He has published in various peer reviewed journals, and he is a reviewer of international journals, including Sustainability; Climate; Science of the Total Environment; and Ecological Economics. His current research interests are in the sustainable use of freshwater for food production. Special attention in his work is given to balance environmental, social, and economic aspects of freshwater use for food production.

Prof. Manfred Max Bergman's main focus is on sustainability and global studies in relation to the UN Sustainable Development Goals, particularly the interdependence of society, business, and government in a globalized world. In pursuing policy-relevant and change-oriented research relating to societal sustainability, he is working on a new social science research approach, entitled Social Transitions Research (STR). He holds the Chair of Social Research and Methodology at the University of Basel, and he is a member of UNESCO NatCom Switzerland and the Sustainable Development Solutions Network (SDSN), a global initiative for the United Nations. He chairs the World Sustainability Forum and currently publishes on corporate sustainability in BRICS countries (Brazil, Russia, India, China and South Africa).

Preface to "Toward a Sustainable Agriculture: Farming Practices and Water Use"

Globally, the agri-food sector has a major role to play in ensuring food security and economic development, especially for lower income countries. It is responsible for producing food that is nutritious and affordable to a growing world population, and for generating income and employment for a large number of people involved along the agri-food chain. As such, the contribution of the agri-food sector extends the mere production of the primary food products at farm level to also include the food processing industry, wholesale and retail. The success with which the agri food sector can meet these responsibilities is very much dependent on the ability of primary producers, at farm level, to supply primary food products that meet the stringent requirements in terms of quantity and quality. Without sufficient supply, the other role-players along the value chain will not be able to meet the demand of consumers. It is also at farm level, where the producers face a variety of risks that negatively affect their ability to produce products that meet the requirements. Environmental risk, exacerbated by climate change, is a major type of risk that negatively affects the ability of primary producers to produce products that meet the above requirements. Changes in the distribution and intensity of rainfall have a major impact on agricultural production, as is well-documented. Agricultural producers have to adapt to find new ways to mitigate the impact of climate change and associated risks on their farming enterprises. Agricultural production is also well documented to contribute significantly to environmental degradation. A new generation of more informed consumers is putting more pressure on retail stores to ensure that the products they buy are produced sustainably. As such, requirements regarding sustainable practices are being specified to all role-players along the value chain, including agricultural producers. Compliance with sustainable production practices is becoming ever more important for being allowed to sell food products, especially in high value markets. Costs associated with compliance place pressure on the budgets of agricultural producers, depressing profit margins. Thin profit margins, often associated with primary agricultural production, mean that producers are not necessarily compensated sufficiently for the risks they have to bare. Ultimately, the farming enterprise still needs to make financial sense for the producer to remain in business. Failing to make ends meet, from a financial perspective, is a major cause for farmers leaving the sector for alternative careers. As such, the ability of the agri-food sector to meets its responsibility in ensuring global food security, and contributing to social and economic development, is increasingly difficult. It is within this context that agricultural producers have to operate, meeting the demand of the agri-food sector to feed growing populations in a sustainable manner while still trying to make ends meet. Agricultural producers are in need of science-based evidence to inform them how to overcome the above challenges. Through this book, a variety of scholars explore risks and potential solutions to inform producers and policy makers regarding practices towards sustainable production, and regarding the sustainable use of freshwater in agricultural production in a context of a changing climate.

Henry Jordaan
Guest Editor

Maximizing Rainfall in Lowland Paddy Rice through Water Depths Control and Alternate Wetting and Drying Irrigation Technique in Southern Taiwan

Victoriano Joseph Pascual and Yu-Min Wang

Abstract: Rainfall along with the use of alternate wetting and drying irrigation technique is proposed to minimize water use and optimize crop yield and water productivity in paddy rice cultivation. A field experiment was conducted to determine the most suitable ponded water depth for reducing paddy rice irrigation. Water treatments of $T2_{cm}$, $T3_{cm}$, $T4_{cm}$, and $T5_{cm}$ were applied weekly from transplanting to early heading through a complete randomized block design with four replications. The results showed that yield loss in $T2_{cm}$ was 3.5 times more than that of $T3_{cm}$ and 14 times more than $T4_{cm}$. The highest irrigation water productivity and total water productivity was produced in $T2_{cm}$, whereas rainwater productivity was greater in $T5_{cm}$. The weekly application of $T4_{cm}$ ponded water depth along with rainfall matched the required crop water and produced the lowest yield reduction and grain production loss, in addition to 20% water saving. Water stress at panicle initiation decreased the daily headed panicle per square meter by 155%, 214%, and 443% in $T4_{cm}$, $T3_{cm}$, and $T2_{cm}$ compared to $T5_{cm}$. However, the decrease of this parameter was followed by total recovery caused by the effective rainfall contribution.

---

1. Introduction

Global agriculture in the 21st century faces the tremendous challenge of providing enough food for a growing population under increasing scarcity of water resources, while minimizing environmental consequences [1,2]. Fresh water for irrigation is becoming increasingly scarce because of population growth, increasing urban and industrial development, and the decreasing availability resulting from pollution and resource depletion [1,3]. The world food security remains largely dependent on irrigated lowland rice which is the main source of rice supply [4]. It provides 75% of the total rice production [1], and consumes more than 50% of total fresh water. By 2025, it is predicted that 15–20 million hectares of irrigated rice field may suffer from physical water scarcity, and the world's farmers should be producing about 60% more rice than at present to meet the food demand of the expected world population [5]. The challenge for sustainable rice production is to decrease the amount of water while maintaining or increasing grain yields [6].

Since the foreseen increase in the supply of rice is constrained by lack of sufficient available water, the most appropriate solution for worldwide water shortage is to make efficient use of agricultural water [7]. China has pioneered various water-saving irrigation technologies to achieve more water-efficient irrigation for agricultural systems [8,9]. Among the various water saving methods, the most widely promoted one for rice is alternate wetting and drying (AWD) irrigation [10,11]. It is reported that AWD practices could reduce both water use and greenhouse gas (GHG) emissions without seriously sacrificing grain yield in rice systems [2]. In AWD, irrigation water is applied to achieve intermittent flooded and non-flooded soil conditions. The frequency of irrigation and duration of non-flooding can be determined by re-irrigating (to achieve flooded conditions) after a fixed number of non-flooded days, when a certain threshold of soil water potential is reached, when the ponded water table level drops to a certain level below the soil surface, when cracks appear on the soil surface, or when plants show visual symptoms of water shortage [8]. Commonly, irrigation is applied to obtain 2–5 cm ponded water depth after a certain number of days (ranging from two to seven) have passed following the disappearance of ponded water [5]. Yield penalty was commonly observed under AWD compared with continuously flooded (CF)-irrigated rice [12]. Generally, however, AWD increased water productivity with respect to total water input because the yield reduction was smaller than the amount of water saved [13]. Tuong et al. [5] assessed the efficiency of AWD compared with CF and found that AWD yielded better than CF in terms of water savings and farm profitability. However, rice is very sensitive to water stress and attempts to reduce water may result in yield reduction and threaten food security. Therefore, efficient irrigation water use requires effective water management during the entirety of the crop production cycle.

Rice is also a very important and valuable crop in Taiwan with a total yield of more than 1.73 million tonnes from 271,077 hectares of land for a production value of TND (Taiwan New Dollar) 41.48 billion (about USD 1.37 billion) in 2014 [14]. There are two cropping seasons for paddy rice in Taiwan. The first crop is cultivated in February and harvested in July, and the second crop is cultivated in August and harvested in December [15]. Taiwan is located in a rainy region with 78% of the rainfall occurring from May to October, with possibilities of rainfall reaching up to 90% in the southern region. Taiwan has an annual average precipitation of 2500 mm, which is higher than the world's average of 834 mm; however, there is still grave water demand, and fresh water for irrigation limits rice production. Apart from rapid urbanization, industrialization, and high irrigation water consumption from the agriculture sector (80%), only a small portion of the water brought by precipitation can be stored over land, as most of the water flows directly into the sea through various rivers in response to steep mountain terrain [16,17], furthermore, this situation is exacerbated by climate change. In 2014, rice production was compromised

2

as a consequence of extended drought forcing the Ministry of Economic Affairs (MOEA) to implement water rationing measures by fallowing approximately 5% of Taiwan's cultivated land [14].

Agriculture water resource scarcity directly affects crop productivity and aggravates food deficit problems for millions of people [18]. In this regard, the maximization of water resources is imperative and the combined effects of rainfall and irrigation management should be addressed together regarding the environment specificities for maximizing water use efficiency and yield per unit of irrigation water applied. Thus, AWD can be optimized and may reduce irrigation cost, increase output, and can particularly be effective in the reproductive and grain filling stages, where rice is more sensitive to water stress. Therefore, the objectives of this research are to apply AWD irrigation to determine the most effective ponded water depth leading to optimum water uptake and low yield losses, while simultaneously maximizing rainfall use alongside scheduled irrigation, and to determine the morphological changes in paddy rice caused by the ponded water treatments. It is expected that the AWD technique can be optimized, through strategic irrigation management while taking full advantage of rainfall occurrences. Such an approach is less documented in areas such as southern Taiwan, hence the reason for the current field experiment.

2. Materials and Methods

2.1. Experimental Site and Trial Design

The experiment was conducted from February to June 2015 in the irrigation experimental field of National Pingtung University of Science and Technology in southern Taiwan, located at 22.39° (N) latitude and 34.95° (E) longitude and 71 m above sea level. The soil type was loamy (27% of sand and 24% of clay) with a wilting point of 15% volume, field capacity 30.5% volume, saturation 42.9% volume, bulk density 1.40 g/cm^3, matric potential 11.09 bar, and hydraulic conductivity 57 mm/hr. The experimental design was a randomized complete block design with four replications and four water treatments. Each plot was 6 m long, 1 m wide, with a total area of 6 m^2, and 0.3 m soil bed height. The spacing between plots and between blocks was 1 m. Ponded water depths were kept constant at 5 cm, 4 cm, 3 cm, and 2 cm representing $T5_{cm}$, $T4_{cm}$, $T3_{cm}$, and $T2_{cm}$.

2.2. Crop Management and Irrigation Management

Twenty-five day old seedlings were obtained from a seed nursery and were manually transplanted on 1 February 2015. Three seedlings were transplanted at hill spacing 25 cm between hills and 20 cm between rows (20 plants m^2). Fertilizer (N:P$_2$O$_5$:K$_2$O) was applied at a ratio of 12:18:12 with a rate of 170 kg/ha at basal, mid tillering, and panicle initiation. Pests were controlled by pesticide application and weeds by frequent manual weeding. Irrigation treatments were applied immediately after transplanting and the irrigation interval was scheduled at seven days. Equation (1) was used to obtain the desired water volume at required depth considering seepage and infiltration in the experimental design and results.

$$IR = A \times h \times 10^3 \tag{1}$$

where IR is the amount of irrigation water (L) for a desired depth above the soil surface, A is the surface area of the plot (m^2), and h is the desired ponded water depth above the soil surface (m). The final irrigation treatment was applied during heading stage on 15 May 2015. Thereafter, the rain was frequent and the crop was subjected to rainy conditions.

2.3. Soil Water Content and Soil Trend Analysis

The soil water content was monitored every two days from one month after transplanting to three weeks before harvest using the gravimetric method. Soil samples were collected using an auger in three different locations within each plot at 25 cm depth. The soil was immediately weighed, and dry weight was obtained after oven drying at 105° C for 24 h. The soil water content per unit was calculated using the following Equation (4).

$$SW = \frac{100 \times (fresh\ weight - dry\ weight)}{dry\ weight} \times \gamma s \tag{2}$$

where SW is the soil water content and γs is the soil bulk density (g/cm^3). The soil water trend was analyzed by determining the soil water content at saturation level, field capacity, wilting point, and stress threshold using Equations (3)–(6) [19].

$$SW_{Sat} = 1000\ (Sat) \times Z_r \tag{3}$$

$$SW_{FC} = 1000\ (FC) \times Z_r \tag{4}$$

$$SW_{WP} = 1000\ (WP) \times Z_r \tag{5}$$

$$SW_{ST} = 1000\ (1 - P)Sat \times Z_r \tag{6}$$

where SW_{Sat}, SW_{FC}, SW_{WP}, and SW_{ST} are soil water content (mm) at saturation, field capacity, wilting point, and stress threshold level, respectively. Sat, FC, and WP are the soil at saturation, field capacity, and wilting point, respectively, in percentage of volume. P is the fraction of water that can be depleted before moisture stress occurs and represents 20% of the saturation for rice crop; Z_r is the sample collection depth (m).

4

2.4. Assessment of Agronomic Parameters

A square meter quadrant which constitutes 20 individual hills was established in the center of each plot to assess plant height and tiller number at panicle initiation and heading stage. Plant height was measured from the base to the tip of the highest leaf while tillers were counted individually per plant. Five hills from each replicate were randomly selected outside the squares for root and biomass per hill assessment at panicle initiation. This was done using an auger 10 cm diameter to remove soil of 20 cm depth from selected hills [20]. A uniform soil volume of 1570 cm^3 was excavated to collect root samples for all treatments. Roots were carefully washed and removed from uprooted plants. Root volume was measured by the water displacement method of putting all the roots in a measuring cylinder and getting the displaced water volume [21]. Root depth was obtained by direct manual measurements of the top root using a ruler against a millimeter paper. Roots dry weight and dry biomass per hill were obtained after oven drying at 70 °C for 24 h.

2.5. Leaf Chlorophyll Content and Relative Water Content

A chlorophyll meter (model SPAD-502, MINOLTA, Osaka, Japan) was used to determine leaf chlorophyll content. Good correlations have been found between the SPAD-502 value and extractable leaves chlorophyll content in several species, although specific calibration is always recommended [22,23]. At panicle initiation and heading stage, 12 hills per plot were selected throughout the diagonals and median, and the 12 uppermost fully expanded leaves were selected from these random hills to analyze the variability of chlorophyll content among treatments with three observations made per leaf. Analysis of leaves sampling patterns done by Chapman and Bareto [24] showed that at least four leaves per plot are needed, with several observations per leaf. Then, the average of these three readings was used to represent the leaf chlorophyll content.

The leaf relative water content (RWC) was calculated from fresh weight (FW), dry weight (DW), and turgid weight (TW) [25].

$$RWC\ (\%) = [(FW - DW)/(TW - DW)] \times 100 \tag{7}$$

2.6. Measuring of Yield and Yield Components

To analyze the heading rate, daily headed panicle numbers was determined in each plot from the appearance of the first panicle until 50% of the farm headed. At harvest, yield components (panicle number per hill, panicle length, and panicle weight, grain number per panicle, grain weight per panicle, and filled grain per panicle) were obtained from inside the square [4]. Panicles were cut at the base, separated from the straw, and the number was determined for each hill. Panicles from each plot were individually measured to determine maximum and minimum length. The range was calculated, and the class interval was obtained by dividing the range by 3 (desired number of classes). Three length classes were determined per plot and panicles were arranged accordingly. Five panicles were randomly picked from each class and the length and weight were measured. The same sampled panicles were individually hand threshed and grain number per panicle was determined. All plants in the squares were harvested, excluding those in edges, for grain yield per unit of area (tha^{-1}) determination. Three samples of harvested grains were randomly picked from each replicate and the dry weight was determined. Grain weight per panicle, and grain yield for unit area was obtained at a constant weight after oven drying at 70 °C for 72 h. The grain yield for unit area was then adjusted at the standard moisture content of 14%. Five samples of 1000 grains were taken from the total grain production of each plot and weighed for the 1000 grain weight determination. Filled spikelets from these samples were separated from unfilled spikelets using a seed blower for 2 mm. The percentage of filled grain was calculated, using mass as the basis, as the ratio of filled grain weight out of the total grain weight multiplied by 100. Fifteen samples were considered per treatment. The dry biomass per hill from the harvested plants was determined after oven drying at 70 °C for 24 h, and the total straw weight (tha^{-1}) was calculated accordingly. The harvest index (HI) was calculated as the ratio of total grain yield out of the total straw yield.

2.7. Water Productivity Assessment

The total water productivity (TWP), irrigation water productivity (IWP) and rain water productivity (RWP) were calculated according to Equations (8)–(10) [26]:

$$TWP = \frac{Y}{TWU} \tag{8}$$

$$IWP = \frac{Y}{IWU} \tag{9}$$

$$RWP = \frac{Y}{RWU} \tag{10}$$

where TWP, IWP, and RWP are the total water (rain + irrigation), irrigation water, and rain water productivity, respectively, expressed in kg·m^{-3}; Y is the grain yield (kg·ha^{-1}), TWU, IWU, and RWU are the total water, irrigation water, and rain water used, respectively, expressed in m^3·ha^{-1}.

Grain production losses were calculated considering the yield in the highest water treatment (T5$_{cm}$) as a reference, and water saving impact was defined as the grain production lost by saving one unit of irrigation water. The water saving impact was obtained by dividing the quantity of grain lost per hectare by the amount of water saved (m^3/ha).

2.8. Data Analysis

The statistical analysis applied on the data includes correlation, and the analysis of variance was done using SPSS 18 software (*PASW Statistics for Windows*, version 18.0.; SPSS Inc.: Chicago, IL, USA, 2009). The significance of the treatment effect was determined using F-test and means were separated through Turkey's test at a 0.05 significance level.

3. Results and Discussion

3.1. Agro-Hydrological Conditions during the Growing Season

The daily maximum temperatures, minimum temperatures, daily rain fall, and crop evapotranspiration (ETc) during the crop production cycle are presented in Figure 1. The weather data were recorded at the National Pingtung University of Science and Technology Agro-Meteorological station. Maximum and minimum temperatures (see Figure 1a) varied from 16.4 to 31.1 °C with a mean value of 25.1 °C, and from 14.1 to 28.8 °C with a mean value of 23.5 °C, respectively. The low values for these two parameters were observed in February while the high values were observed in June. February was recorded as the driest month during the crop cycle.

Daily rainfall (see Figure 1b) ranged from 0 to 81.3 mm with monthly recorded values (February, March, April, May, and June 2015) of 5.2, 0.9, 11.2, 229.2, and 30.1 mm, respectively. Rainfall was more frequent during the month of May compared to other months and coincided with the final stages of panicle initiation and throughout heading. The ETc was obtained by multiplying the reference crop evapotranspiration (ETo) per adjusted crop coefficient (Kc) [14]. Crop evapotranspiration varied along the production cycle and ranged from 1.33 to 3.12 mm/day with the lowest observed value in February (vegetative stage) and the highest value observed in April (panicle initiation). From panicle initiation up to the onset of harvest the crop water demand was above 2 mm/day.

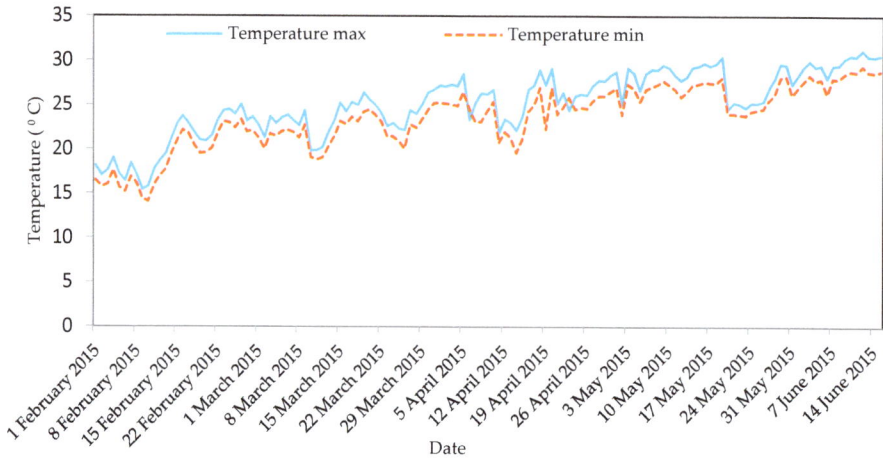

(a)

Figure 1. *Cont.*

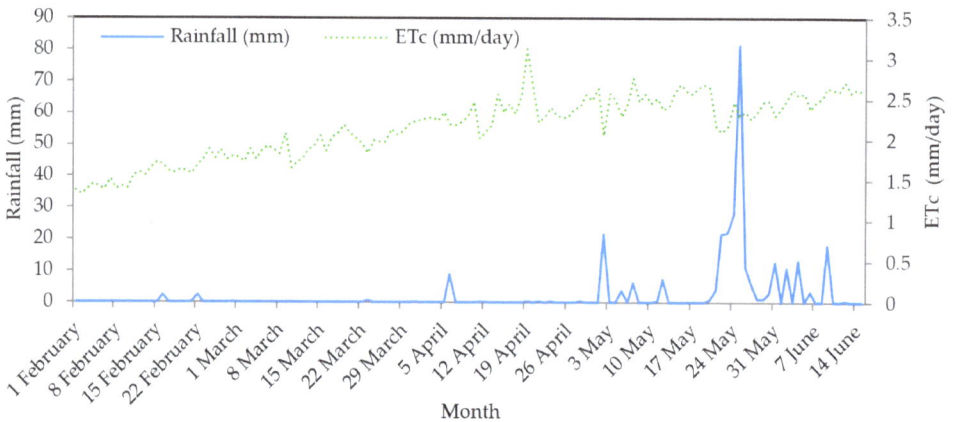

(b)

Figure 1. (**a**) Daily maximum and minimum temperatures; (**b**) Daily rainfall and crop evapotranspiration during the crop production cycle.

According to the growth stages, 62 m^3/ha of rain was recorded during the vegetative growth stage (February–March), 509 m^3/ha during panicle initiation (April–12 May 2015), and 2197 m^3/ha from heading (12 May 2015) to harvest (16 June 2015). During the vegetative stage rainfall represented 3.87%, 2.58% 1.93%, and 1.55% of irrigation water applied in treatments T2$_{cm}$, T3$_{cm}$, T4$_{cm}$, and T5$_{cm}$, respectively. From panicle initiation to heading, it represented 42.41%, 28.27%, 21.21%, and 16.96% of the same treatments. Plants were almost entirely grown under irrigation at the vegetative stage; on the contrary, they were subjected to both irrigation and rainfall during panicle initiation and almost exclusively grown under rain-fed condition from heading to harvest. The highest rainfall contribution throughout the crop production cycle occurred from heading to harvest.

3.2. Soil Water Content of Different Water Treatments

The soil water trend analysis was based on the soil stress threshold and was recorded from the vegetative to heading stage during the crop production cycle (see Figure 2). Throughout the vegetative stage, the soil water content reached its maximum every two days after irrigation and then a sharp decline occurred until the next irrigation. The highest water treatment ($T5_{cm}$) produced the highest soil water content throughout the crop production cycle. Soil water content varied according to irrigation treatments, but was usually between soil stress thresholds and/or above soil saturation level for all water treatments during the vegetative stage. The soil water content for $T2_{cm}$ produced the lowest values throughout the growing cycle, but was never below the soil stress threshold during the vegetative stage, whereas that of $T4_{cm}$ and $T3_{cm}$ was frequently between highest and lowest water treatments at this time. At panicle initiation, the soil water content for $T2_{cm}$ fell below the soil stress threshold level six times, with 79.4 mm recorded as the lowest value. Soil water content for $T3_{cm}$ fell below the soil stress threshold level once (84.8 mm), however, low values of 87.3, 86.3, and 87.5 mm were also recorded. From 18 May 2015 onwards, rainfall was frequent and irrigation was suspended. The soil water content at this time was closer to saturation levels in all treatments. Previous studies [27,28] have confirmed that the critical stages for rice sensitivity to water stress are during panicle initiation, anthesis, and grain filling. Boonjung and Fukai [29] highlighted that plants which suffer mild stress during early panicle development stage suffered yield reduction of around 30% due to reduction in the number of spikelets per panicle. Water stress was observed during the high crop water requirement period (panicle initiation), with $T2_{cm}$ being significantly affected compared to the other treatments.

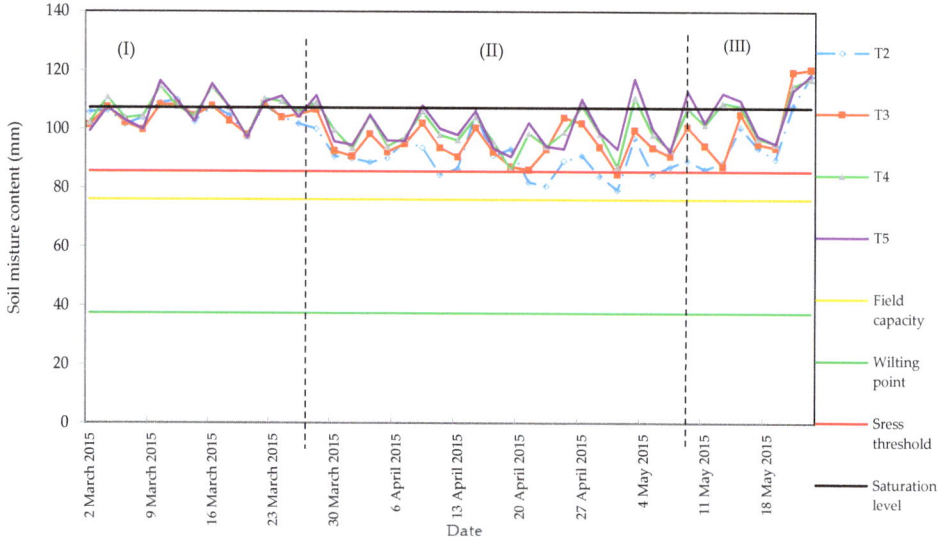

Figure 2. Soil water content at (I) vegetative, (II) panicle initiation, and (III) early heading stage.

3.3. Crop Growth

Growth parameters of plant height and tiller numbers presented in Table 1 show that plant height was significantly affected by water treatments at panicle initiation and heading stage. Under water stress, plants reduced evapotranspiration which led to decreases in photosynthesis which in turn induced the decrease of chlorophyll, height, and tiller number [13]. Reddy et al. [30] concluded that drought stress induced a decline in net photosynthesis and reduced growth rate. Low plant heights were notable in lower water treatments, with $T2_{cm}$ and $T3_{cm}$ showing significant height differences compared to $T5_{cm}$ at the panicle initiation. At heading, the lowest plant height was recorded in $T2_{cm}$, while comparable height was seen among $T5_{cm}$, $T4_{cm}$, and $T3_{cm}$. Water stress in $T3_{cm}$ was not as severe as that of $T2_{cm}$, hence the reason $T3_{cm}$ was able to produce comparable height to $T5_{cm}$ at heading. Water restrictions at panicle initiation decreased average plant height by 8.35%, 4.21%, and 2.50%, while at heading height was reduced by 4%, 2.9%, and 2.2% in $T2_{cm}$, $T3_{cm}$, and $T4_{cm}$, respectively. A high correlation ($R^2 = 0.92$ and $R^2 = 0.93$, respectively) was found between plant height and irrigation water application during panicle initiation and heading (see Figure 3a,b). It is well known that water restriction may retard plant growth and reduce plant height, however, plants subjected to slight water stress conditions during the panicle initiation stage recovered faster under well water conditions. Lilley and Fukai [31] demonstrated that severe water deficit suspended apical development until re-watering occurred, while mild water deficit reduced the

10

rate of apical development. Kima et al. [32] confirmed that plants recovered from the effects of water stress that occurred during vegetative stage and performed as well as the highest water treatment at heading stage. The extent of recovery due to re-watering strongly depends on pre-drought intensity and duration [33].

No significant differences were observed for tiller numbers among water treatments, however, the smallest tillers and the lowest tiller numbers were observed in $T2_{cm}$. Nguyen et al. [34], in comparing various water saving systems in rice, found no significant difference in tiller number among water treatments and suggested that tillering was less sensitive than other characteristics, such as plant height and leaf area. Akram et al. [35] also noted that in all growth stages, tiller number per hill of different rice cultivars was not significantly affected by soil moisture stress. Results show correlation ($R^2 = 0.78$) at panicle initiation and ($R^2 = 0.82$) at heading, thereby revealing significant correlations between irrigation water and tiller number (see Figure 3c,d).

Table 1. Effects of water treatments on plant height and tiller numbers.

Treatments	Panicle Initiation		Heading	
	Plant Height	Tiller Numbers	Plant Height (cm)	Tiller Numbers
T5	71.88 [a]*	14.64	86.68 [a]	19.72
T4	70.12 [ab]	13.93	84.77 [ab]	19.82
T3	68.97 [b]	14.16	84.17 [ab]	19.32
T2	68.65 [b]	13.28	83.32 [b]	18.75
P	*	ns	*	ns

* = mean with columns not followed by the same letter indicate a significant difference at the $p < 0.05$ level as determined by Tukey's test; ns = not significantly different.

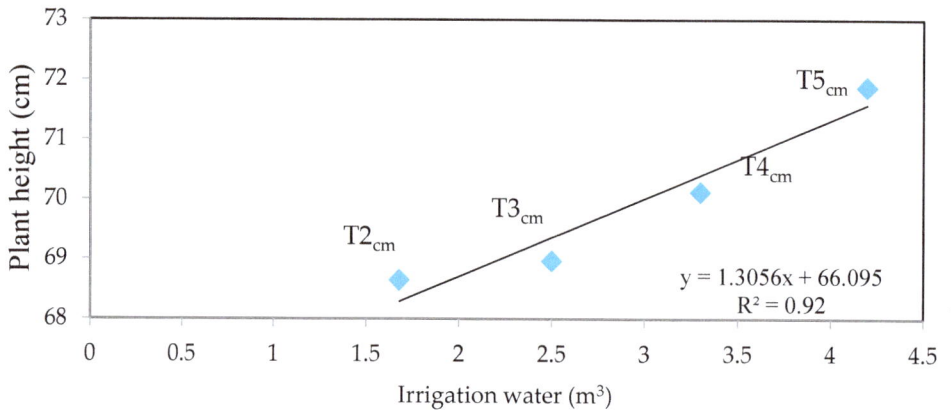

(a)

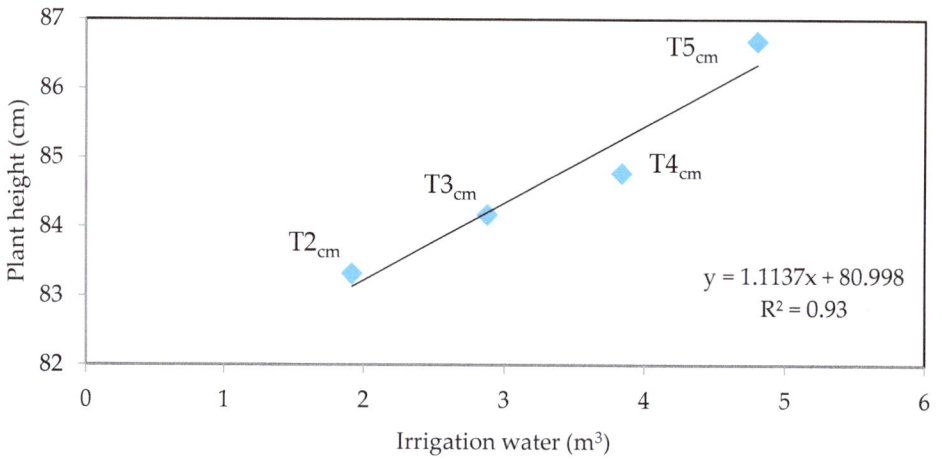

(b)

Figure 3. *Cont.*

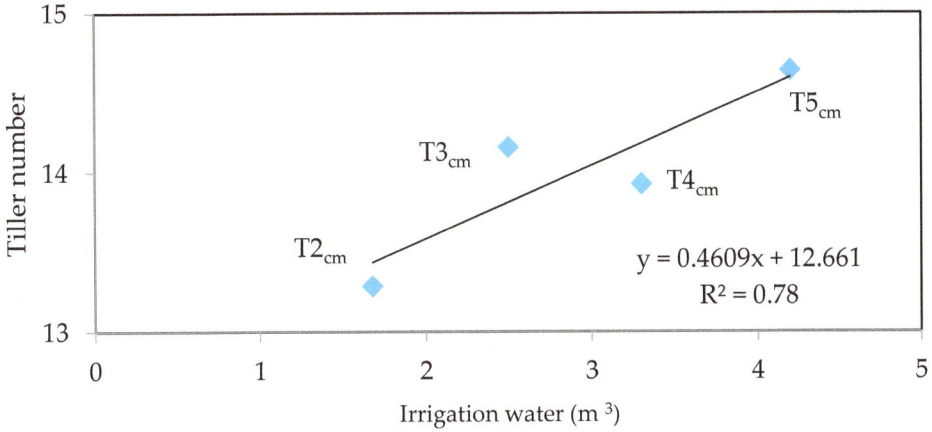

(c)

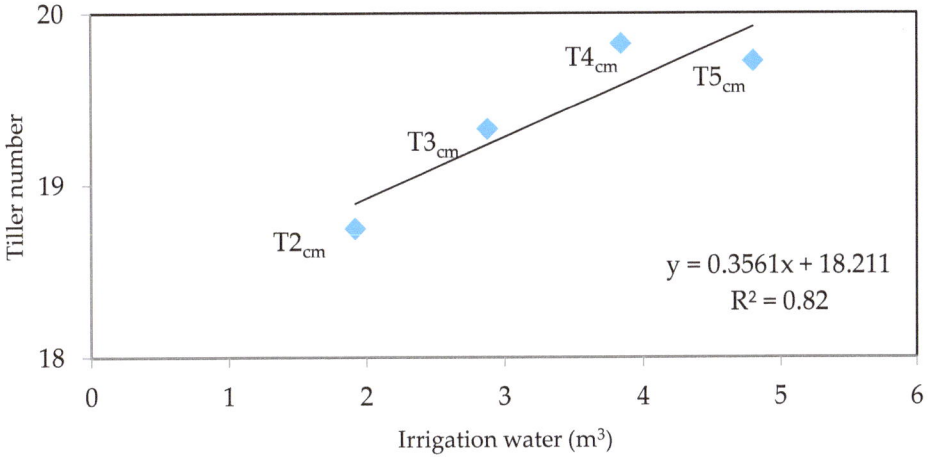

(d)

Figure 3. Relationship between plant height and irrigation water at panicle initiation (**a**) and heading (**b**) and between tiller numbers and irrigation water at panicle initiation (**c**) and heading (**d**).

3.4. Dry Biomass, Root Dry Weight, Root Depth, and Root Volume

The results of root parameters and dry biomass are presented in Table 2. The highest values for dry biomass, root volume, and root dry weight was produced in T5$_{cm}$. No significant differences were observed between dry biomass and root volume for T3$_{cm}$ and T4$_{cm}$, however, significant differences were observed between

13

T5$_{cm}$ and T2$_{cm}$. Blackwell et al. and Turner et al. [36,37] observed that biomass production decreases with decreasing water availability. In addition, dry biomass accumulation is one of the main growth factors of rice and large root dry weight matter as high root activity implies strong water and nutrient absorption capacity, which tend to favor high grain production [38,39]. Lilley and Fukai [31] explained that rice cultivars differ in physiological response to water deficit that is associated with differences in water extraction capability. Furthermore, the response of different plants to water stress is much more complex, and various mechanisms are adopted by plants when they encounter drought stress at various growth stages [40]. The results also indicate that no significant differences were observed for root depths for all treatment. Ascha et al. [41] highlighted that plants become adapted to water deficiency through the possession of a pronounced root system, which maximizes water capture and allows access to water depth. Kima et al. [32] evaluated various water treatments for water use efficiency under saturated soil culture and noted that lower root depth and root weight values were produced in higher water treatments. They concluded that such results may be explained by the effects of hydraulic head pressure, which may affect infiltration rate. Further explanation revealed that, in higher water treatments, water depth on the soil surface may lead to an infiltration rate that matches in time with water uptake, and hence the availability of soil water may not reach a critical point for the crop to develop a deeper root system as an adaptation measure [32]. Root dry weight and root volume were significantly higher for T5$_{cm}$ when compared to T2$_{cm}$, but no significant differences were found among T2$_{cm}$, T3$_{cm}$, and T4$_{cm}$. Further observation showed that roots were thicker and fuller in 0–10 cm soil in T5$_{cm}$ when compared to the other water treatments. In addition, healthy roots were observed in all treatments. Root health may be attributed to repeated wetting and drying practiced under AWD. Ndiiri et al. [16] explained that of the several factors that contribute to high nitrogen availability and high nitrogen usage efficiency under system of rice intensification (SRI) management practices, the repeated wetting and drying process may have the greatest influence; moreover, lack of aeration of soil affects not only root health and functions, but also the populations of beneficial organisms that contribute to plant nutrition and health. There is also evidence that phosphorus solubilization and availability are increased by alternate wetting and drying [32].

Table 2. Effect of water treatment on dry biomass, root dry weight, root depth, and root volume at panicle initiation.

Treatments	Dry Biomass (g/hill)	Root Dry Weight (g/hill)	Root Depth (cm)	Root Volume (cm^3)
T5	28.80 [a]*	14.48 [a]	17.10	21.00 [a]
T4	27.71 [ab]	13.04 [ab]	16.89	20.30 [ab]
T3	28.85 [a]	10.94 [ab]	16.44	17.40 [ab]
T2	22.71 [b]	9.33 [b]	15.56	15.15 [b]
P	*	*	ns	*

* = mean with columns not followed by the same letter indicate significant difference at the $p < 0.05$ level as determined by Tukey's test; ns = not significantly different.

3.5. Leaf Chlorophyll Content and Relative Water Content

The effect of water treatments on leaf chlorophyll content and leaf relative water content (RWC) at panicle initiation is presented in Table 3. It is well established that AWD exposes crops to temporary water stress during the drying cycles and that plants adapt to water stress by stomatal closure, change in leaf turgor, and chlorophyll fluorescence. Leaf greenness is an indicator of a plant's health, and it may be affected by both leaf nitrogen content and water stress [4]. Chlorophyll content and RWC was highest in (T5$_{cm}$), with no significant differences observed among the other treatments. Cha-Um et al. [42] explained that, in evaluating water deficit stress in four *indica* rice genotypes, RWC in the flag leaf was positively correlated with total chlorophyll; moreover, total chlorophyll and total carotenoids in all rice cultivars were drastically degraded when subjected to severe water stress. However, the degradation percentage of the pigments would recover and greatly improve after re-watering. It was also noted that water use efficiency in rice subjected to water deficit declined significantly. Furthermore, Zhang et al. [43] highlighted that under an alternate wetting and severe soil drying regime (WSD), cytokinin levels were reduced when compared to conventional irrigation and alternate wetting and moderate soil drying (WMS). The explanation in support stated that changes in hormones in leaves under different treatments were closely associated with those of the photosynthetic rate, with a high correlation observed between hormone content and photosynthetic rate. There was no significant difference in chlorophyll content and RWC at the heading stage. RWC value was higher during the heading stage compared with the panicle initiation stage, since at this time the soil water content was usually higher due to frequent rain fall. Akram et al. [35] and Lafitte [44] explained that reduction of leaf RWC was related to soil water content, especially in water deficit stress cultivars.

Table 3. Chlorophyll content and leaf relative water content subjected to water treatments.

Treatments	Panicle Initiation		Heading	
	Chlorophyll Content	RWC	Chlorophyll Content	RWC
T5	46.85 [a*]	70.43 [a]	44.50	85.77
T4	43.92 [b]	62.57 [b]	43.43	84.90
T3	45.34 [ab]	65.02 [ab]	44.16	77.93
T2	43.61 [b]	60.05 [b]	43.72	85.15
P	*	*	ns	ns

* = mean with columns not followed by the same letter indicate significant difference at the $p < 0.05$ level as determined by Tukey's test; ns = not significantly different.

3.6. Effect of Water Treatment on Yield Components and Grain Yield

Daily headed panicle and panicle emergence were affected by water treatments (see Figure 4). Panicle numbers in $T5_{cm}$ were significantly higher, and emergence was faster compared with other treatments. When compared to $T5_{cm}$, panicle reduction rate per square meter was 155%, 214%, and 443% in, $T4_{cm}$, $T3_{cm}$, and $T2_{cm}$, respectively. The high occurrences of water stress in $T2_{cm}$ at panicle initiation caused significant declines in headed panicle per m^2, showing that water restriction affected the number of reproductive tillers. By delaying plant growth, water stress during panicle initiation delayed the heading rate, which decreased the panicle number per hill. Akram et al. [30] explained that severe soil moisture stress at panicle initiation was more destructive to panicle number per hill, panicle length, panicle dry weight, shoot dry weight, and total grains per panicle, irrespective of the cultivars, resulting in a drastic decrease in per hectare paddy yield. O'Toole and Moya [45] highlighted that water deficit at any growth stage may reduce such conditions based on the magnitude of the reduction which is dependent on the severity, timing, and duration.

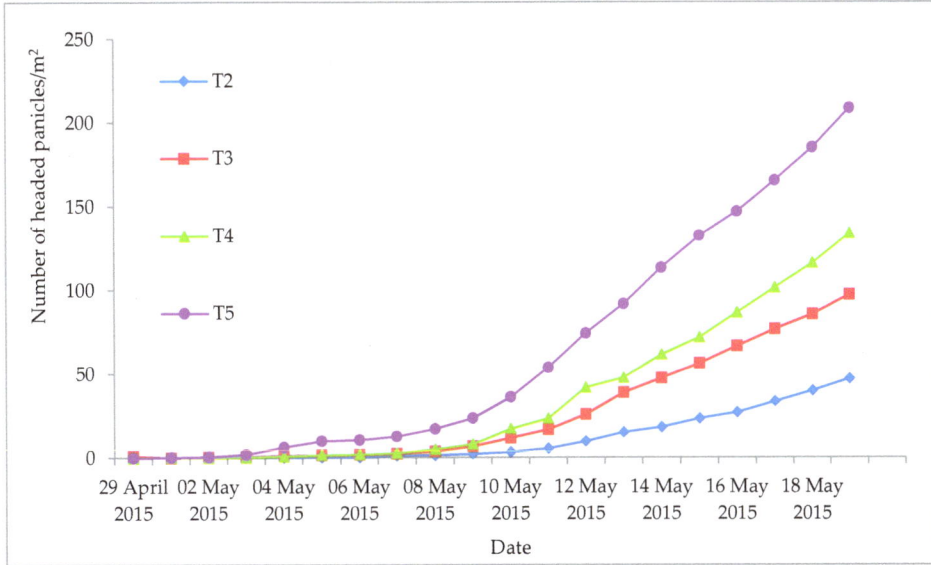

Figure 4. Effects of water treatments on daily headed panicle.

The results presented in Table 4 show that no significant differences were produced for average panicle number per hill, average panicle length, and average panicle weight; however, average panicle weight decreased with the lowest water treatment. Davatgar et al. [46] explained that mild water stress at mid tillering affects assimilates translocation from most plant parts to the panicles, via altering source sink relationships. The reduction in leaf cell expansion decreased the sink strength for vegetative growth and lessened the competition with panicle growth assimilates. Even though water stress occurred during the panicle initiation stage, it is well established that tillering and panicle initiation may occur simultaneously, and probably contributed to the translocation of assimilates during this time. The effect of assimilates being translocated from plant parts may be one of the reasons for the yielding of comparative results. From heading to harvest, a total of 2768 m^3/ha of rainfall was registered which may have also contributed towards overcoming the effects of water stress which occurred during early panicle initiation in some treatments. Kima et al. [4] explained that high rainfall occurring from heading to harvest allowed crops to overcome the effects of water stress experienced during vegetative stage, leading to recovery of yield components such as panicle number and grain number per panicle. Turner et al. [37] also suggested that the reduction of yield component by water stress is dependent on severity, timing, and duration; furthermore, previously established literature [33] has indicted that the response of

17

different plants to water stress is much more complex and that various mechanism are adopted by plants when they encounter drought stress at various growth stages.

Table 4. Water treatments effects on panicle number, panicle weight, and panicle length at harvest.

Treatments	Average Panicle Number per Hill	Average Panicle Weight (g)	Average Panicle Length (cm)
T5	16.87	2.04	24.52
T4	15.74	2.01	25.35
T3	16.32	1.91	24.97
T2	15.44	1.71	24.65
P	ns	ns	ns

Ns = not significantly different at $p < 0.05$ level as determined by Tukey's test.

The values in Table 5 show that water treatments significantly affected the average grain number per panicle, grain weight per panicle, grain filling rate, and 1000 grain weight. The lowest values of these parameters were observed in $T2_{cm}$. There were no significant differences observed for average grain number per panicle between $T5_{cm}$ and $T2_{cm}$, however, water stress significantly affected grain weight per panicle in $T2_{cm}$ and grain filling rate in $T2_{cm}$ and $T3_{cm}$. Grain weight per panicle was reduced by 32% in $T2_{cm}$, while unfilled grain percentage was 18.7%, 20.9%, 25.1%, and 31.1% for $T5_{cm}$, $T4_{cm}$, $T3_{cm}$, and $T2_{cm}$, respectively. For 1000 grain weight, 17.5% of the weight was lost in $T2_{cm}$. Since there was a delay in heading, and panicle initiation occurred at the same time with flowering, water stress greatly affected the flowering stage. This might be because water stress slowed down carbohydrate synthesis and/or weakened the sink strength at reproductive stages and aborted fertilized ovaries [47]. As a result, this may have induced spikelet sterility or grain filling delay, leading to high unfilled grain percentage in $T2_{cm}$.

Table 5. Water amounts effects on grain number per panicle, grain weight per panicle, grain filling rate, and 1000 grain weight.

Treatments	Grain Number per Panicle	Grain Weight per Panicle (g)	Grain Filling Rate %	1000-Grain Weight (g)
T5	109.20 ab*	1.91 a	81.31 a	15.99 a
T4	112.83 a	1.99 a	79.15 a	15.70 a
T3	110.85 a	1.80 ab	74.91 b	14.80 a
T2	107.09 b	1.67 b	68.88 c	13.60 b
P	*	*	*	*

* = mean with columns not followed by the same letter indicate significant difference at the $p < 0.05$ level as determined by Tukey's test.

The results of straw weight, grain yield, and harvest index in Table 6 show that these parameters were affected by water treatments. Grain yield in T2$_{cm}$ was significantly reduced by water stress which occurred during panicle initiation. Even though water stress occurred simultaneously in T3$_{cm}$, it was not as severe, hence T3$_{cm}$ was able to produce comparable crop yield to T5$_{cm}$ and T4$_{cm}$. The yield loss in T2$_{cm}$ was 3.5 times more than that of T3$_{cm}$ and 14 times more than that of T4$_{cm}$. The lowest yield reduction was observed in T4$_{cm}$ at 1.57%.

Table 6. Effect of water on straw weight, grain yield, harvest index, yield loss, and yield reduction.

Treatments	Straw Weight (ton/ha)	Grain Yield (ton/ha)	Harvest Index (HI)	Yield Loss (kg/ha)	Yield Reduction %
T5	12.09 a*	5.74 a	0.48 a	—	—
T4	11.77 ab	5.65 a	0.48 a	90	1.57
T3	11.71 b	5.35 a	0.46 a	390	6.79
T2	10.98 c	4.48 b	0.41 b	1260	21.95
P	*	*	*	—	—

* = mean with columns not followed by the same letter indicates significant difference at the $p < 0.05$ level by Tukey's test.

3.7. Water Use Efficiency

Table 7 highlights the results in terms of the amount of rainfall, irrigation, and water use efficiency. Cumulative rainfall recorded from transplanting to harvest represented 35%, 43%, 58%, and 87% of the gross irrigation water applied in T5$_{cm}$, T4$_{cm}$, T3$_{cm}$, and T2$_{cm}$, respectively. The highest rainwater productivity was achieved in T5$_{cm}$ (2.07 kg/m^3), and then gradually decreased to the lowest water treatment T2$_{cm}$ (1.62 kg/m^3). The highest total water productivity, 0.75 kg/m^3, and irrigation water productivity, 1.40 kg/m^3, were observed in the lowest water treatment T2$_{cm}$. The lowest grain production loss (0.06 kg) was observed in T4$_{cm}$, indicating that 0.06 kg of grain was lost for saving 1 m^3 of water. Therefore, based on lowest yield reduction (1.57%) and grain production loss (0.06 kg), the weekly application of 4 cm ponded water depth led to optimal water productivity and a water saving of 20%, and appeared suitable and beneficial for rice crops. In conducting similar research with an emphasis on saturated soil culture, Yao et al. [13] explained that 3 cm soil saturation provided the best results based on lowest yield reduction, lowest grain production loss, and water savings. However, other variables such as environmental conditions and crop variety may also determine the outcome of such results.

Table 7. Effect of treatments on water use efficiency.

Treatments	Rain (m^3/ha)	Irrigation (m^3/ha)	TWP (kg/m^3)	RWP (kg/m^3)	IWP (kg/m^3)	Water Savings (m^3/ha)	Irrigation Water Savings (%)	Water Saving Impact (kg/m^3)
T5	2768	8000	0.53	2.07	0.72	—	—	—
T4	2768	6400	0.62	2.04	0.88	1600	20	0.06
T3	2768	4800	0.71	1.93	1.11	3200	40	0.12
T2	2768	3200	0.75	1.62	1.40	4800	60	0.26

TWP = total water productivity; IWP = irrigation water productivity; RWP = rain water productivity.

4. Conclusions

The challenges to sustaining rice productivity are presently increasing, as there is greater scarcity of water and more competition for water resources. This study has shown that AWD can be optimized through efficient irrigation management and rainfall maximization, thereby concurrently achieving the dual goals of increasing grain production and reducing the water requirements for irrigated paddy rice. Rainfall induced favorable watered conditions that rose and kept the soil moisture content above the soil stress threshold during the final stages of panicle initiation and throughout the heading stage. The weekly application of T4$_{cm}$ ponded water depth from transplanting to heading produced the lowest yield reduction and grain production loss while having no significant impact on yield loss compared to the highest water treatment, and is therefore suitable for increasing irrigation water productivity. On the contrary, plants exposed to T2$_{cm}$ ponded water depth were more vulnerable under soil water stress and showed a reduction in yield components and overall grain yield. Likewise, the number of daily headed panicle per square meter was most affected by T2$_{cm}$ (443%) when compared to T5$_{cm}$. The weekly application of T5$_{cm}$ ponded water depth from transplanting to heading increased the rain water productivity but induced low irrigation water productivity; on the contrary, irrigation water productivity and total water productivity was greater in

T2$_{cm}$. Since rain water use is free of cost, excess use of irrigation water during the dry season in (T5$_{cm}$) appeared costly and non-beneficial. By applying T4$_{cm}$ ponded water depth, and synchronizing the high crop water demand period with the onset of the rainy season, AWD technique efficiency can noticeably be improved. For sustainable rice production and agriculture in general, different methods and technologies for minimizing water usage are explored; however, it has been demonstrated repeatedly that high rice yields can be achieved under non-flooding conditions, and AWD is only one of several techniques which offer opportunities to raise rice production using less water. In this context, combining irrigation and maximizing rainfall can reduce rice farmers' need for irrigation water, enhance grain production, and assist in alleviating food and water shortages in rice producing countries with similar environmental conditions. Finally, water is not an easily fungible resource, and the hydrological dynamics across time and place need to be taken into account. The results presented merit further exploration taking into account additional water depths as soil hydrological condition, timing of irrigation, crop variety, and agronomic attributes may also affect crop yield.

Author Contributions: The authors are both well conversant with the content of the manuscript and have agreed to the sequence of the authorship. Victoriano Joseph Pascual conducted the field work and wrote the manuscript. Yu-Min Wang supervised the field work, and provided oversight for the analysis of data and editing of the manuscript.

Conflicts of Interest: The authors declare no conflict of interest.

References

1. Bouman, B. A conceptual framework for the improvement of crop water productivity at different spatial scales. *Agric. Syst.* **2007**, *93*, 43–60. [CrossRef]
2. Linquist, B.A.; Anders, M.M.; Adviento-Borbe, M.A.A.; Chaney, R.L.; Nalley, L.L.; Da Rosa, E.F.; Kessel, C. Reducing greenhouse gas emissions, water use, and grain arsenic levels in rice systems. *Glob. Chang. Biol.* **2015**, *21*, 407–417. [CrossRef] [PubMed]
3. Belder, P.; Bouman, B.; Cabangon, R.; Guoan, L.; Quilang, E.; Yuanhua, L.; Spiertz, J.; Tuong, T. Effect of water-saving irrigation on rice yield and water use in typical lowland conditions in Asia. *Agric. Water Manag.* **2004**, *65*, 193–210. [CrossRef]
4. Kima, A.S.; Chung, W.G.; Wang, Y.-M.; Traoré, S. Evaluating water depths for high water productivity in irrigated lowland rice field by employing alternate wetting and drying technique under tropical climate conditions, southern Taiwan. *Paddy Water Environ.* **2014**, *13*, 379–389. [CrossRef]
5. Fageria, N. Yield physiology of rice. *J. Plant Nutr.* **2007**, *30*, 843–879. [CrossRef]
6. Yang, J.; Zhang, J. Crop management techniques to enhance harvest index in rice. *J. Exp. Bot.* **2010**, *61*, 3177–3189. [CrossRef] [PubMed]
7. Wang, Y.-M.; Namaona, W.; Traore, S.; Zhang, Z.-C. Seasonal temperature-based models for reference evapotranspiration estimation under semi-arid condition of Malawi. *Afr. J. Agric. Res.* **2009**, *4*, 878–886.

8. Peng, S.; Bouman, B. Prospects for genetic improvement to increase lowland rice yields with less water and nitrogen. In *Scale and Complexity in Plant Systems Research: Gene-Plant-Crop Relations*; Spiertz, J.H.J., Struik, P.C., Van Laar, H.H., Eds.; Springer: Dordrecht, The Netherlands, 2007; pp. 251–266.

9. Zhang, H.; Xue, Y.; Wang, Z.; Yang, J.; Zhang, J. An alternate wetting and moderate soil drying regime improves root and shoot growth in rice. *Crop Sci.* **2009**, *49*, 2246–2260. [CrossRef]

10. Cabangon, R.; Castillo, E.; Tuong, T. Chlorophyll meter-based nitrogen management of rice grown under alternate wetting and drying irrigation. *Field Crops Res.* **2011**, *121*, 136–146. [CrossRef]

11. Tuong, P.; Bouman, B.; Mortimer, M. More rice, less water—Integrated approaches for increasing water productivity in irrigated rice-based systems in Asia. *Plant Prod Sci.* **2005**, *8*, 231–241. [CrossRef]

12. Bouman, B.; Tuong, T.P. Field water management to save water and increase its productivity in irrigated lowland rice. *Agric. Water Manag.* **2001**, *49*, 11–30. [CrossRef]

13. Yao, F.; Huang, J.; Cui, K.; Nie, L.; Xiang, J.; Liu, X.; Wu, W.; Chen, M.; Peng, S. Agronomic performance of high-yielding rice variety grown under alternate wetting and drying irrigation. *Field Crops Res.* **2012**, *126*, 16–22. [CrossRef]

14. Council of Agriculture, Executive Yuan R.O.C. The Mid-term Agricultural Program of the Council of Agriculture, 2013. Council of Agriculture, Executive Yuan R.O.C. Available online: http://eng.coa.gov.tw/theme_data.php?theme=eng_policies&id=9 (accessed on 21 December 2016).

15. Liou, Y.-A.; Liu, H.-L.; Wang, T.-S.; Chou, C.-H. Vanishing ponds and regional water resources in Taoyuan, Taiwan. *Terr. Atmos. Ocean.* **2015**, *26*, 161–168. [CrossRef]

16. Hsiao, T.; O'Toole, J.; Tomar, V. Water stress as a constraint to crop production in the tropics. In *International Rice Research Institute, Priorities for Alleviating Soil-Related Constraints to Food Production in the Tropics*; Drosdoff, M., Zandstra, H., Rockwood, W.G., Eds.; International Rice Research Institiute: Manila, Philippines, and New York State College of Agriculture and Life Sciences, Cornell University, 1980; pp. 339–369.

17. Lee, J.-L.; Huang, W.-C. Impact of climate change on the irrigation water requirement in northern Taiwan. *Water* **2014**, *6*, 3339–3361. [CrossRef]

18. Traore, S.; Wang, Y.-M.; Kan, C.-E.; Kerh, T.; Leu, J.M. A mixture neural methodology for computing rice consumptive water requirements in Fada N'Gourma region, eastern Burkina Faso. *Paddy Water Environ.* **2010**, *8*, 165–173. [CrossRef]

19. Allen, R.G.; Pereira, L.S.; Raes, D.; Smith, M. *Crop Evaporation–Guidelines for Computing Crop Water Requirements–Fao Irrigation and Drainage Paper 56*; Food and Agriculture Organization of the United Nations: Rome, Italy, 1998.

20. Kawata, S.; Katano, M. On the direction of the crown root growth of rice plants. *Jpn. Crop Soc.* **1976**, *45*, 471–483. [CrossRef]

21. Ndiiri, J.; Mati, B.; Home, P.; Odongo, B.; Uphoff, N. Comparison of water savings of paddy rice under system of rice intensification (sri) growing rice in Mwea, Kenya. *Int. J. Curr. Res. Rev.* **2012**, *4*, 63–73.

22. Marenco, R.; Antezana-Vera, S.; Nascimento, H. Relationship between specific leaf area, leaf thickness, leaf water content and spad-502 readings in six amazonian tree species. *Photosynthetica* **2009**, *47*, 184–190. [CrossRef]
23. Markwell, J.; Osterman, J.C.; Mitchell, J.L. Calibration of the minolta spad-502 leaf chlorophyll meter. *Photosynth. Res.* **1995**, *46*, 467–472. [CrossRef] [PubMed]
24. Chapman, S.C.; Barreto, H.J. Using a chlorophyll meter to estimate specific leaf nitrogen of tropical maize during vegetative growth. *Agron. J.* **1997**, *89*, 557–562. [CrossRef]
25. Bonnet, M.; Camares, O.; Veisseire, P. Effects of zinc and influence of acremonium lolii on growth parameters, chlorophyll a fluorescence and antioxidant enzyme activities of ryegrass (lolium perenne l. Cv apollo). *J. Exp. Bot.* **2000**, *51*, 945–953. [CrossRef] [PubMed]
26. Pereira, L.S.; Cordery, I.; Iacovides, I. Improved indicators of water use performance and productivity for sustainable water conservation and saving. *Agric. Water Manag.* **2012**, *108*, 39–51. [CrossRef]
27. Tao, H.; Brueck, H.; Dittert, K.; Kreye, C.; Lin, S.; Sattelmacher, B. Growth and yield formation of rice (oryza sativa) in the water-saving ground cover rice production system (gcrps). *Field Crops Res.* **2006**, *95*, 1–12. [CrossRef]
28. Jian-Chang, Y.; Kai, L.; Zhang, S.-F.; Xue-Ming, W.; Zhi-Qin, W.; Li-Jun, L. Hormones in rice spikelets in responses to water stress during meiosis. *Acta Agron. Sin.* **2008**, *34*, 111–118.
29. Boonjung, H.; Fukai, S. Effects of soil water deficit at different growth stages on rice growth and yield under upland conditions. 2. Phenology, biomass production and yield. *Field Crops Res.* **1996**, *48*, 47–55. [CrossRef]
30. Reddy, A.R.; Chaitanya, K.V.; Vivekanandan, M. Drought-induced responses of photosynthesis and antioxidant metabolism in higher plants. *J. Plant Physiol.* **2004**, *161*, 1189–1202. [CrossRef]
31. Lilley, J.; Fukai, S. Effect of timing and severity of water deficit on four diverse rice cultivars iii. Phenological development, crop growth and grain yield. *Field Crops Res.* **1994**, *37*, 225–234. [CrossRef]
32. Kima, A.S.; Chung, W.G.; Wang, Y.-M. Improving irrigated lowland rice water use efficiency under saturated soil culture for adoption in tropical climate conditions. *Water* **2014**, *6*, 2830–2846. [CrossRef]
33. Xu, Z.; Zhou, G.; Shimizu, H. Plant responses to drought and rewatering. *Plant Signal. Behav.* **2010**, *5*, 649–654. [CrossRef] [PubMed]
34. Nguyen, H.; Fischer, K.; Fukai, S. Physiological responses to various water saving systems in rice. *Field Crops Res.* **2009**, *112*, 189–198. [CrossRef]
35. Akram, H.; Ali, A.; Sattar, A.; Rehman, H.; Bibi, A. Impact of water deficit stress on various physiological and agronomic traits of three basmati rice (oryza sativa l.) cultivars. *J. Anim. Plant. Sci.* **2013**, *23*, 1415–1423.
36. Blackwell, J.; Meyer, W.; Smith, R. Growth and yield of rice under sprinkler irrigation on a free-draining soil. *Aust. J. Exp. Agric.* **1985**, *25*, 636–641. [CrossRef]

37. Turner, N.C.; O'Toole, J.C.; Cruz, R.; Namuco, O.; Ahmad, S. Responses of seven diverse rice cultivars to water deficits. Stress development, canopy temperature, leaf rolling and growth. *Field Crops Res.* **1986**, *13*, 257–271. [CrossRef]

38. Kato, Y.; Okami, M. Root growth dynamics and stomatal behaviour of rice (oryza sativa l.) grown under aerobic and flooded conditions. *Field Crops Res.* **2010**, *117*, 9–17. [CrossRef]

39. Mishra, A.; Salokhe, V. Flooding stress: The effects of planting pattern and water regime on root morphology, physiology and grain yield of rice. *J. Agron.Crop Sci.* **2010**, *196*, 368–378. [CrossRef]

40. Jones, H. What Is Water Use Efficiency. In *Water Use Efficiency in Plant Biology*; Bacon, M., Ed.; Wiley-Blackwell: Hoboken, NJ, USA, 2009; pp. 24–47.

41. Ascha, F.; Dingkuhn, M.; Sow, A.; Audebert, A. Drought-induced changes in rooting patterns and assimilate partitioning between root and shoot in upland rice. *Field Crops Res.* **2005**, *93*, 223–236. [CrossRef]

42. Cha-Um, S.; Yooyongwech, S.; Supaibulwatana, K. Water deficit stress in the reproductive stage of four indica rice (oryza sativa l.) genotypes. *Pak. J. Bot.* **2010**, *42*, 3387–3399.

43. Zhang, H.; Chen, T.; Wang, Z.; Yang, J.; Zhang, J. Involvement of cytokinins in the grain filling of rice under alternate wetting and drying irrigation. *J. Exp. Bot.* **2010**, *61*, 3719–3733. [CrossRef] [PubMed]

44. Lafitte, R. Relationship between leaf relative water content during reproductive stage water deficit and grain formation in rice. *Field Crops Res.* **2002**, *76*, 165–174. [CrossRef]

45. O'Toole, J.; Moya, T. Water deficits and yield in upland rice. *Field Crops Res.* **1981**, *4*, 247–259. [CrossRef]

46. Davatgar, N.; Neishabouri, M.; Sepaskhah, A.; Soltani, A. Physiological and morphological responses of rice (oryza sativa l.) to varying water stress management strategies. *Int. J. Plant Prod.* **2012**, *3*, 19–32.

47. Rahman, M.; Islam, M.; Islam, M. Effect of water stress at different growth stages on yield and yield contributing characters of transplanted aman rice. *Pak. J. Biol. Sci.* **2002**, *5*, 169–172.

Sustainability of Irrigating Winter Wheat in the East of England—Adaptation at What Cost?

Daniel El Chami and André Daccache

Abstract: Climate change is "the challenge" of our times and for the next upcoming decades. The biggest impacts are likely to affect the sustainability of agricultural and food systems; both highly vulnerable to continuously changing climatic patterns. Wheat is a strategic crop for food security. It is widely grown worldwide as a rainfed (unirrigated) crop; but the latest research shows that recent world wheat price increases and increasing weather variability are making supplemental irrigation marginally profitable. The proposed study combines the outputs from a general circulation model (GCM), the Food and Agriculture Organization of the United Nations (FAO) crop growth model (AquaCrop), and economic modelling to assess the sustainability of irrigated wheat production compared to rainfed crop production both under current climate conditions and in the future under different climate scenarios. The AquaCrop model has been calibrated and validated for winter wheat grown on a sandy loam soil in the East of England (Bedfordshire). Long-term observed climate data (1970–2006) in Cambridge (Cambridgeshire) were used to validate the projected climate data from the GCM. Structural characteristics of the case study were representative of a typical farm of the area, and irrigation costs and wheat prices for the economic model were calculated assuming current market prices. In the longer term, a sensitivity analysis was used to assess the expected variations due to the increase in world wheat prices and the energy costs involved. Results of the study show that the impacts of climate change on winter wheat grown in the East of England would be a reduction in the rainfed yield (between -5.4% and -32.9%) and that the projected economic losses from rainfed winter wheat production would be expected to range between -24.3% and -36.0%. Irrigation, which does not seem to be an economic option under the current climate conditions, could be a future adaptation measure for yield increase (3.9–6.1 t·ha^{-1}) and to improve the financial appraisal of irrigation investment, which would raise between 41 and 519 £·ha^{-1}. However, negative externalities are increasing pressures on water and air resources, for example, an increase of the irrigation water requirements between 25.0% an 39.1% and global warming potential increases between 2.5% and 21.5%. Finally, the study suggests further research to incorporate a life cycle assessment model into the framework for an integrated and comprehensive approach for sustainability assessment of wheat in particular and agricultural systems in general.

25

1. Introduction

The sustainability of world agricultural and food systems is highly endangered by the impacts of climate warming due to the high vulnerability of these systems to continuously changing climatic patterns [1] (p. 976). At the global scale, climate impacts are expected to further decrease world crop yields [2]. However, heterogeneous results are expected at regional levels due to local variations in agro-climatic conditions [3].

Wheat is the most important crop grown worldwide for food provision and feed for livestock [4,5]. Given its high adaptability to different climate conditions, it is grown under very diverse agro-climates extending from Russia to the tropics and sub-tropics [6]. In many areas, wheat is grown as a rainfed crop, but irrigation occurs in some areas such as sub-tropics [7].

Literature on the impacts of climate change on wheat production is abundant but contradictory. Parry et al. [2] adopted statistical analyses to derive agro-climatic regional yield transfer functions from site-level results under different Special Report on Emission Scenarios (SRES) and showed that wheat, amongst other cereal crops (wheat, rice maize, and soybean), is subject to potential yield changes at global levels, which would expose the global food security to high risks and result in security consequences [8]. Contrarily, Wilcox and Makowski [9] contradicted the previous results showing in a meta-analysis that the effects of high CO_2 concentrations would outweigh the effects of increasing temperature and the decline in precipitation leading to increasing yields depending on the geographical location. However, Supit et al. [10] used outputs from three general circulation models (GCMs) and a crop growth monitoring system in combination with a weather generator to demonstrate that crops planted in autumn and winter, such as winter wheat in Europe, may benefit from the increasing CO_2 concentration in the short run, but if the CO_2 increase lessens or stagnates, yield reductions may occur after 2050.

At localized levels, literature assessing the impacts of climate change on wheat crop tackled single aspects (e.g., impact on yield and/or water use) and results are site-specific [11–13]. However, Falloon and Betts [14] recommended an integrated approach to deal with climate change research, which is lacking so far.

Therefore, the aim of this study is to assess the sustainability of winter wheat production at the farm level, adopting irrigation for adaptation to climate change in a typical temperate climate in the East of England (UK). It will adopt an integrated modelling approach to estimate potential trade-offs between water savings, energy consumption (greenhouse gas (GHG) emissions), and economic benefits under current and future climate scenarios. This integrated approach makes a significant contribution to the carbon accounting of crop production in general and the impacts of intensification through irrigation in particular, and it could be easily replicated with different case studies and other crops.

2. Material and Methods

The study was divided into the following stages:

1. Defining a typical wheat-growing farm for modelling assessment.
2. Selecting the baseline climate data from a local weather station and downscaling the data according to different scenarios.
3. Quantifying the irrigation water requirements (depths applied) under current climate conditions and estimating the yield response and yield benefits from irrigation.
4. Assessing sustainability of the rainfed and irrigated winter wheat using a selection of financial, environmental, and social indicators under current climate conditions and future scenarios.
5. Undertaking a sensitivity analysis to assess the effects of variation in costs and market prices.

2.1. Selection of a "Typical" Farm

The case study farm was selected to reflect the regional farm characteristics in the East of England (Table 1). Therefore, we assumed that the farm was 200 ha, practicing rotational agriculture with winter wheat occupying 50 ha annually. The on-farm irrigation system was a hose reel fitted with either a raingun or boom, the most common method of irrigation in the UK according to Department for Environment, Food and Rural Affairs (DEFRA) [15], using an all year abstraction license from a nearby river and a diesel pump. We modelled irrigation needs assuming a deep uniform sandy loam soil, with a soil depth of 4 m and a total available water of 120 mm/m, as irrigation in England is more likely to be used on the lighter, droughtier soils (e.g., [16]); we considered, however, a deep uniform silty clay loam soil with a depth of 4 m and a total available water of 210 mm/m, which is a heavier soil given that most wheat is currently grown on heavier soils.

Table 1. Wheat production summary statistics for England and the East of England for 2011 [17,18].

Indicator	England	East of England	East of England/England (%)
Farmed area ($\times 10^6$ ha) *	8.89	1.38	15.5
Total number of farms *	53,090	8147	15.0
Wheat area ($\times 10^6$ ha) *	1.79	0.50	28.0
Wheat yield (t·ha^{-1})	7.73	7.21	93.3
Average farm size (ha) *	153.3	195.4	127.5
Wheat production (Mt)	13.8	3.6	26.1
Wheat output (Million £)	1984.64	573.51	28.9
Total crop output (Million £)	7724.42	1,979.58	25.6

* Data relates to 2010.

2.2. Climate Data and Climate Scenarios

The observed climate dataset used in this study was daily data (1970 to 2006) from a meteorological weather station located at Cambridge, Cambridgeshire (52.24° N, 0.10° W). Data included rainfall, reference evapotranspiration (ET_0), and maximum and minimum temperature for the historical baseline period (Figure 1).

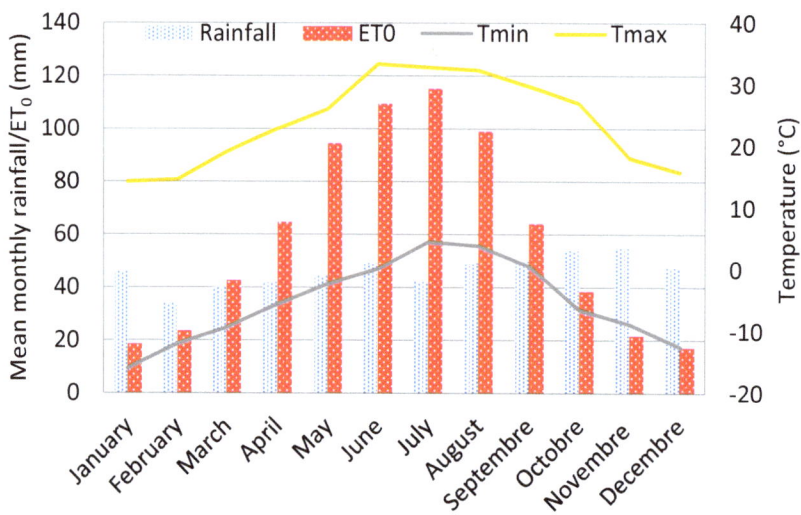

Figure 1. Monthly observed climate dataset at Cambridge.

To generate the future weather dataset, a LARS-WG stochastic weather generator was used [19,20] to produce daily weather from GCM outputs at a single site. The HadCM3 was chosen for this study since it was one of the major GCMs used in the Intergovernmental Panel on Climate Change (IPCC) Third Assessment Report [21], and has been widely used in the literature for climate impacts assessment (e.g., [22,23]).

The emissions scenarios used are those developed by the IPCC [24] and known as SRES (Special Report on Emission Scenarios), in which each scenario combines two sets of divergent tendencies: one set varies between strong economic values and strong environmental values, the other set between increasing globalization and increasing regionalization [25]. The scenarios are commonly known as A1 (economic-global), B1 (environmental-global), A2 (economic-regional), and B2 (environmental-regional). For this research A1 and B1 emissions scenarios were used; both are characterized by a rapid economic growth and a global population that reaches nine billion in 2050 and then gradually declines. The A1 scenario characterizes a future with energy technology balanced between fossil and non-fossil fuel, whilst the B1 scenario reflects global efforts to control GHG emissions through the introduction of clean and resource-efficient technologies (Table 2).

Table 2. Characteristics of climate change scenarios for the 2050s (A1 and B1) according to selected scenarios of the Intergovernmental Panel on Climate Change (IPCC) Special Report on Emission Scenarios (SRES) [25].

Characteristics	IPCC Scenarios	
	Scenario A1	Scenario B1
Population growth (billion)	8.7 (Low)	8.7 (Low)
World GDP (10^{12} 1990 USD/yr)	164–187 (Very High)	136 (High)
Energy use (10^{18} J/yr)	1213–1431 (Very High)	813 (Low)
Global CO_2 emissions (GtC)	13.5 (Medium)	4.2 (Low)
Land-use changes	Low	High
Resource availability	Medium–High	Low
Technological change	High	Medium
Change favoring	Coal-Balanced Non-Fossils	Efficiency and Dematerialization

2.3. Irrigation Water Requirements (IWR) and Yield Response

The historical baseline and downscaled future climate datasets for Cambridge were used as inputs for the crop growth modelling. The Food and Agriculture Organization of the United Nations (FAO) crop model AquaCrop was used to simulate potential yield as a function of water consumption. AquaCrop has been previously tested under similar climate conditions (e.g., [26,27]). Further, El Chami et al. [26] have calibrated and validated AquaCrop for winter wheat crop in the East of England against experimental yield data obtained from the Broadbalk wheat experiment at Rothamsted Experimental Research Station (Harpenden, UK).

In this study, the same irrigation schedule adopted by El Chami et al. [28] was set (irrigation period between 1 April and 1 June applying 25 mm at a 50 mm soil moisture deficit), to maintain a small deficit in the rootzone to maximize the effective use for rainfall. Indeed, under typical UK climate conditions, irrigation on winter wheat is not generally needed before April, and should stop before the beginning of June with the initiation of flowering [29,30]. Furthermore, experimental studies in the East of England in the 1990s showed that irrigation on cereals after flowering would increase the risk of logging [31].

2.4. Sustainability Assessment

To date, considerable efforts have been made to identify appropriate indicators for agricultural sustainability [32,33], because indicators are one of two basic approaches to sustainability assessment [30]. Indicators can be divided into the multi-dimensional components of sustainability (economic, environmental, and social).

To evaluate the economic efficiency of winter wheat production, a cost-benefit analysis (CBA) was a key component of the integrated framework for the financial investment appraisal (FIA) of different options. The net present values (NPVs) were calculated to assess the economic viability over the life cycle of the project. Other economic indicators were also selected for this same purpose: the internal rate of returns (IRRs) to measure the capacity of the net revenues to remunerate the investment cost, and the benefit-cost ratios (BCRs) to summarize the relative size of the present benefits with respect to the present costs.

To estimate the economic model parameters, we consider a six-year average wheat price (2007–2012) as reported by Home-Grown Cereal Authority (HGCA) [34] for milling wheat. The production costs of the typical rainfed winter wheat farm in the UK are based on an integration of figures for the 2012 harvest year from Agro Business Consultants (ABC) [35] and from survey data from Nix [36]. The abstraction charge calculations are based on Environment Agency charges for 2013/14 [37]. The capital cost of the irrigation system calculations are based on updated market figures for similar equipment from a major local equipment supplier (Briggs Irrigation UK), whilst the variable elements of irrigation costs (labor, fuel, machinery) are based on an updated analysis of detailed irrigation costs by Morris et al. [38].

The social cost of carbon (SCC) was also calculated. It is defined as the estimated price of the damages caused by each additional ton of carbon dioxide (CO_2) released into the atmosphere [39], accounting also for the other GHG using carbon dioxide equivalents. The social cost of irrigation systems was included in the appraisal converting the volume of diesel used in the operation into global warming potential (GWP) [40] and multiplying it by a ten-year average of non-tradable prices of carbon obtained from Department of Energy & Climate Change (DECC) [41]; the average of the baseline was from 2008 to 2017 and the projected average of the future scenarios was from 2041 to 2050.

A selection of water related indicators and CO_2 emission indicators were used to assess the environmental effectiveness of wheat production on water and air resources. The first indicator is the GWP (tCO_2e per hectare) defined as a unit of measurement that allows the effect of different GHGs and other factors to be compared using CO_2 as a standard unit of reference. Three water related indicators were selected according to the data available to calculate them: (1) the surface water withdrawals (WWs), which in this case is the volume of irrigation water applied on the farm per unit area. This has been introduced by FAO as a key environmental integrity indicator of water resources [32] and it could be a good indicator to compare the water savings per hectare between different on-farm practices. (2) The irrigation use efficiency (IUE) used in the literature as an indicator to maximize water productivity and sustainably allocate resources [42] and defined in our case as the ratio of yield increment ($t·ha^{-1}$) due to the irrigation water requirement ($m^3·ha^{-1}$). (3) The added value of water (AVW), which is the extra benefit of irrigation generated per unit volume of water which shows in economic terms how water contributes to the production value.

To assess the social dimension in this research, we adopted the food security indicator classified under the safety indicators described by the FAO [30], which could support shocks and increase human well-being. In this case, food security was measured through the yield increase (YI). It is also to be noted that income increase could also be considered as a social indicator, because the extra money generated could be spent on farmers' well-being.

2.5. Sensitivity Analysis

Whilst analysts ascribe the current drop in oil prices to political drivers and objectives [43] similar to those described by Stevens [44,45] behind the non-regulation of the supply curve to reach a Pareto optimal price, oil prices are, however, expected to continue their upward trend. Previous estimates suggest that the price of oil is expected to rise by up to 60% by 2035 with prices of USD 250 per barrel forecasted [46]. This affects farmers, because fuel price is one of the major factors influencing operating costs for irrigated crops [47]. Further, the cereal price index has been generally increasing since 2000 and prices by 2022 are projected to be between 12% and 27% above those of the previous decade [48–51]. Therefore, a sensitivity analysis was carried out to find out how the added value of winter wheat would respond to price fluctuations and variations in the total costs of production.

3. Results and Discussion

This section first describes the results of simulated yields and the irrigation water requirements, then the sustainability indicators will be assessed for rainfed and irrigated production in different soils for the baseline climate and under different scenarios. Finally, the sensitivity analysis will estimate the variation of the indicators for wheat price change and for oil prices.

3.1. Yields and IWR Estimates

Under current climate conditions (baseline scenario), the results show that rainfed winter wheat crops grown on lighter soils (sandy loam) produce 14% less yield than heavier ones (silty clay loam). The results partially agree with He et al. [52] who showed that heavier soils have higher water use (WU), hence, wheat grown on lighter soil textures have good growth and high yields providing there is sufficient summer rains to replenish soil water.

In the future, rainfed yield is expected to be negatively affected by climate warming (between −5.4% and −32.9% depending on the scenario and the soil type) (Figure 2), which confirms the findings of Semenov and Shewry [53] who warned that a warming climate would have negative impacts on UK wheat; it endorses as well the results of other studies in similar climate conditions expecting negative impacts of climate change on winter wheat (e.g., [3,54]). Conversely, the results contradict the conclusions of Richter and Semenov [55], who predicted a yield increase for winter wheat of 15%–23% by the 2050s. This could be due, according to Kersebaum and Nendel [54], to site-specific conditions, as they noted a difference in simulated yield even within regions as site conditions had a strong influence on crop growth.

It should be noted that the yield variability of rainfed winter wheat under climate change scenarios is higher than the baseline scenarios (±2.4 for sandy loam soil and ±1.2 for silty clay loam soil), which has been abundantly stressed in the literature (e.g., [56,57]). However, irrigation (as discussed in the introduction) could be an efficient technique for adaptation to climate change as it reduces uncertainties and increases yields for a future food insecure population (Table 3). However, this adaptation measure comes at a cost. For the same irrigation schedule, a higher IWR is required in the future (25%–39% higher) compared to the current conditions (Figure 3), which is in line with the conclusions of Weatherhead and Knox [58] who predicted a future rise in irrigation water requirements in England and Wales.

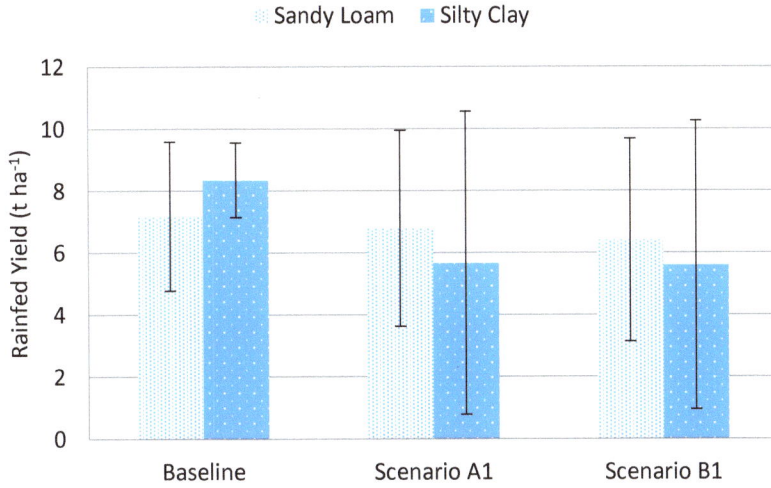

Figure 2. Simulated yield (t·ha^{-1}) at Cambridge for baseline and selected IPCC SRES scenarios [25].

Table 3. Rainfed and irrigated yields for different soil types and under baseline and selected IPCC SRES scenarios [25].

Soil Type	Sandy Loam Soil		Silty Clay Loam Soil	
Yield (t·ha^{-1})	Rainfed	Irrigated	Rainfed	Irrigated
Baseline	7.2 ± 2.4	8.7 ± 0.9	8.4 ± 1.2	9.2 ± 0.8
Scenario A1	6.8 ± 3.2	10.7 ± 0.7	5.7 ± 4.9	11.8 ± 0.7
Scenario B1	6.4 ± 3.3	10.5 ± 1.0	5.6 ± 4.7	11.7 ± 0.7

Figure 3. Irrigation water requirements (IWRs) for the same irrigation schedule under baseline and selected IPCC SRES scenarios [25].

3.2. Sustainability Assessment

Currently, under the baseline scenario rainfed winter wheat produced in sandy loam soils generate 23% more benefits than in silty clay soils; further, irrigation is not a beneficial option to farmers and generates environmental externalities in terms of CO_2 emissions and added value of water (Table 4).

Table 4. Sustainability assessment of agricultural practices for different soil types, under baseline and selected IPCC SRES scenarios [25].

Soil	Sandy Loam Soil						Silty Clay Loam Soil					
Scenario	Baseline		Scenario A1		Scenario B1		Baseline		Scenario A1		Scenario B1	
	Boom	Raingun	Boom	Raingun	Boom	Raingun	Boom	Raingun	Boom	Raingun	Boom	Raingun
FIA	−93.9	−184.4	260.1	104.5	294.1	144.9	−354.5	−535.4	429.4	67.4	417.9	41.0
NPV	10278.6	9770.2	12858.5	12108.3	12585.5	11868.5	10432.0	9171.0	13570.6	11748.0	13416.6	11516.5
IRR	83%	83%	101%	99%	99%	99%	84%	84%	107%	105%	105%	105%
BCR	6.7	4.3	7.6	4.5	7.5	4.5	5.5	2.6	6.1	2.6	5.9	2.5
WW	680	680	880	880	850	850	1380	1380	1850	1850	1920	1920
IUE	0.013	0.013	0.012	0.012	0.012	0.012	0.007	0.007	0.006	0.006	0.006	0.006
AVW	0.38	0.24	0.52	0.34	0.50	0.33	0.14	0.01	0.24	0.04	0.22	0.02
GWP	0.1	0.4	0.1	0.5	0.1	0.5	0.2	0.8	0.3	1.2	0.3	1.1
YI	1.6	1.6	3.9	3.9	4.1	4.1	0.9	0.9	6.1	6.1	6.1	6.1

FIA = Financial Investment Appraisal; NPV = Net Present Value; IRR = Internal Rate of Return; BCR = Benefit-Cost Ratio; WW = Water Withdrawal; IUE = Irrigation Use Efficiency; AVW = Added Value of Water; GWP = Global Warming Potential; YI = Yield Increase.

Under climate change scenarios, the yield reduction for rainfed winter wheat would cause a benefit loss to farmers between -24.3% and -36.0% on a sandy loam soil. On a silty clay loam soil farmers would not generate any benefit. These figures agree with El Chami et al. [28] who suggested that currently the yield benefits do not justify the investment in new irrigation systems for winter wheat grown in the UK. However, using existing on-farm unused equipment available between April and June could be beneficial.

Therefore, irrigation improves the farm economic performances and the added value of water is higher (Table 4), but this might increase CO_2 emissions between 2.5% and 21.5% according to the SRES scenario [25] selected, the soil type, and the irrigation system adopted. This range is not as wide as the results of Niero et al. [59] who assessed the life cycle of spring barley in Denmark under seven alternative future scenarios and found a GWP variation between -31% and 50% compared to baseline.

In general, irrigating winter wheat grown in heavier soils might generate higher incomes, but would require more water (two folds higher) and therefore the CO_2 emissions would be higher (Table 4). Further, results under all scenarios show that raingun systems are not a sustainable on-farm option to adopt; the global warming potential associated with the use of a raingun system is between 40% and 92.2% higher compared to the boom systems and the AVW is between 5.4% and 95.4% lower for rainguns (Table 4). Even though the capital cost of a raingun is cheaper than a boom, the variable costs and the social cost of carbon are relatively high, which makes them an expensive option with high pollution related impacts.

Under climate change scenarios, investing in irrigation systems becomes noticeably profitable and necessary to increase food security (YI: 3.9 to 6.1 t·ha^{-1}). The higher benefits are observed on a silty clay loam soil and for the high emission scenario (Scenario A1). However, the highest economic benefits consume the highest amount of water (WW 1850–1920 m^3·ha^{-1}) and generate a lower AVW accompanied with the highest GWP, which could be two times higher in silty clay loam soils than in lighter soils (Table 4).

3.3. Sensitivity Analysis

According to the sustainability assessment, the most sustainable practice in the future to grow winter wheat would be on lighter soils, and given that irrigation activity would likely be affected by the components assessed in the sensitivity analysis (e.g., market wheat price and oil price), the sensitivity analysis was performed on the sandy loam soil and considered the AVW, which is a good indicator to show in economic terms how water contributes to the production value.

In general, the results of this analysis showed that the AVW is more sensitive to market price fluctuations than to variations in oil prices (Table 5). However, it is well noted that sensitivity of the AVW when using a hose reel fitted with boom (Table 5a) is less elastic than when using a raingun (Table 5b), hence a change in oil price would have higher impacts on the AVW when using a raingun.

Table 5. Sensitivity of Added Value of Water ($£·m^{-3}$) to market price of wheat ($£·t^{-1}$) and diesel price ($£·L^{-1}$) when using **(a)** a hose reel fitted with boom or **(b)** a raingun.

				(a)					
				Sandy Loam Soil—Boom					
	Baseline			**Scenario A1**			**Scenario B1**		
	−30%	165.9	+30%	−30%	165.9	+30%	−30%	165.9	+30%
−60%	−0.24	0.40	1.04	−0.06	0.54	1.15	−0.09	0.53	1.14
0.7	−0.26	**0.38**	1.02	−0.09	**0.52**	1.13	−0.11	**0.50**	1.12
+60%	−0.29	0.35	0.99	−0.11	0.50	1.10	−0.14	0.48	1.10
				(b)					
				Sandy Loam Soil—Raingun					
	Baseline			**Scenario A1**			**Scenario B1**		
	−30%	165.9	+30%	−30%	165.9	+30%	−30%	165.9	+30%
−60%	−0.30	0.34	0.98	−0.16	0.44	1.05	−0.19	0.43	1.04
0.7	−0.40	**0.24**	0.88	−0.26	**0.34**	0.95	−0.29	**0.33**	0.94
+60%	−0.50	0.14	0.78	−0.36	0.24	0.85	−0.36	0.23	0.84

Bold values = Central values

4. Methodological Limitations

Certainly, the integration of different modelling approaches adopted in this study has numerous limitations. The crop growth modelling was based on one GCM (HadCM3), two scenarios (A1 and B1), one time-slice (2050s), two irrigation systems (hose reel fitted with boom and with raingun), and two soil types (Sandy loam and silty clay loam). A more detailed assessment would need to consider the entire range of projections and different soil types and irrigation systems in order to better quantify modelling errors and uncertainties [60].

Even though the irrigation systems used are both sprinklers with the same theoretical efficiency, the on-site application efficiency might vary from one system to another and abstraction efficiency might also be different. The study did not account for application and abstraction efficiency, which may have reduced the accuracy to estimate the applied IWRs and the ground and surface WW to assess the water related indicators.

Finally, the research accounted only for the GHG emissions generated from irrigation pumping and did not include the total emissions of the production generated through the life cycle of the wheat crop. This might have led to an underestimation in the results for the generated indicators.

5. Conclusions

This study confirmed that climate change would have negative impacts on winter wheat production in the East of England. These impacts are site-specific and highly depend on the agro-climatic conditions of the farm. Climate change impacts are also extremely dependent on soil types. The rainfed yields would be reduced by between -5.4% and -32.9% according to the soil type.

In economic terms, rainfed winter wheat under climate change would cost farmers between -24.3% and -36.0% of the benefit margin. However, irrigation could be a beneficial adaptation measure for farmers to increase future yields (YI between 3.1 and 6.1 t·ha^{-1}) and reduce yield uncertainties. It could generate economic benefits that vary depending on the irrigation systems selected (FIA between 41 and 519 £·ha^{-1}), the SRES scenario [25] adopted, and the soil type.

Irrigation might generate environmental externalities and might increase pressure on water and air resources as the irrigation water requirements increase between 25.0% and 39.1% compared to the baseline scenario, and the global warming potential increases between 2.5% and 21.5%. However, the selection of irrigation systems with low energy consumption would limit the environmental impacts to a minimum. As such, hose reels fitted with boom compared to hose reels fitted with raingun showed to be a more sustainable option which should be considered in the future in order to increase food security.

Finally, this research has attempted to integrate different modelling approaches to assess the sustainability of a wheat production system in a humid climate. However, for a comprehensive framework to be adopted and replicated in different agro-climatic conditions, different crops and a bigger range of SRES scenarios, a life cycle assessment model should be incorporated into the framework for wheat in particular and agricultural systems in general.

Acknowledgments: The authors gratefully acknowledge the South African national research foundation (NRF) for funding the outputs of this research through the Knowledge, Interchange and Collaboration programme (KIC). Furthermore, the editorial assistance and support provided by Mrs Liesl van der Westhuizen are significantly appreciated.

Author Contributions: Both authors conceived the idea and agreed on material and methods. André Daccache simulated climate change scenarios and Daniel El Chami did all the other modelling and wrote the paper which has been reviewed by André Daccache in its final version. Both authors have read and approved the final manuscript.

Conflicts of Interest: The authors declare no conflict of interest.

References

1. Parry, M.L.; Canziani, O.F.; Palutikof, J.P.; van der Linden, P.J.; Hanson, C.E. (Eds.) *Climate Change 2007: Impacts, Adaptation and Vulnerability. Contribution of Working Group II to*

the Fourth Assessment Report of the Intergovernmental Panel on Climate Change; Cambridge University Press: Cambridge, UK, 2007.

2. Parry, M.L.; Rosenzweig, C.; Iglesias, A.; Livermore, M.; Fischer, G. Effects of climate change on global food production under SRES emissions and socio-economic scenarios. *Glob. Environ. Chang* **2004**, *14*, 53–67. [CrossRef]

3. Lehmann, N.; Finger, R.; Klein, T.; Calanca P and Walter, A. Adapting crop management practices to climate change: Modelling optimal solutions at the field scale. *Agric. Syst.* **2013**, *117*, 55–65. [CrossRef]

4. Shewry, P.R. Wheat. *J. Exp. Bot.* **2009**, *60*, 1537–1553. [CrossRef] [PubMed]

5. Curtis, B.C.; Rajaram, S.; Gómez Macpherson, H. (Eds.) *Bread, Wheat: Improvement and Production*; Plant Production and Protection Series 30; Food and Agriculture Organization of the United Nations (FAO): Rome, Italy, 2002.

6. Feldman, M. Wheats. In *Evolution of Crop Plants*; Smartt, J., Simmonds, N.W., Eds.; Longman Scientific and Technical: Harlow, UK, 1995; pp. 185–192.

7. Food and Agriculture Organization of the United Nations (FAO). *Crop Water Information: Wheat*; FAO Water Development and Management Unit, 2013. Available online: http://www.fao.org/nr/water/cropinfo_wheat.html (accessed on 15 December 2016).

8. Barnett, J. Security and climate change. *Glob. Environ. Chang.* **2003**, *13*, 7–17. [CrossRef]

9. Wilcox, J.; Makowski, D. A meta-analysis of the predicted effects of climate change on wheat yields using simulation studies. *Field Crop Res.* **2014**, *156*, 180–190. [CrossRef]

10. Supit, I.; van Diepen, C.A.; de Wit, A.J.W.; Wolf, J.; Kabat, P.; Baruth, B.; Ludwig, F. Assessing climate change effects on European crop yields using the crop growth monitoring system and a weather generator. *Agric. Forest Meteorol.* **2012**, *164*, 96–111. [CrossRef]

11. Guo, R.; Lin, Z.; Mo, X.; Yang, C. Responses of crop yield and water use efficiency to climate change in the North China Plain. *Agric. Water Manag.* **2010**, *97*, 1185–1194. [CrossRef]

12. Luo, Q.; Bellotti, W.; Williams, M.; Bryan, B. Potential impact of climate change on wheat yield in South Australia. *Agric. Forest Meteorol.* **2005**, *132*, 273–285. [CrossRef]

13. Eitzinger, J.; Štastná, M.; Žalud, Z.; Dubrovský, M. A simulation study of the effect of soil water balance and water stress on winter wheat production under different climate change scenarios. *Agric. Water Manag.* **2003**, *61*, 195–217. [CrossRef]

14. Falloon, P.; Betts, R. Climate impacts on European agriculture and water management in the context of adaptation and mitigation – the importance of an integrated approach. *Sci. Total Environ.* **2010**, *408*, 5667–5687. [CrossRef] [PubMed]

15. Department for Environment, Food and Rural Affairs (DEFRA). Water Usage in Agriculture and Horticulture. Results from the Farm Business Survey 2009/10 and the Irrigation Survey, 2011. Swarm Hub. Available online: https://goo.gl/KE1KG9 (accessed on 8 April 2013).

16. Daccache, A.; Weatherhead, E.K.; Stalham, M.A.; Knox, J.W. Impacts of climate change on irrigated potato production in a humid climate. *Agric. Forest Meteorol.* **2011**, *151*, 1641–1653. [CrossRef]

17. Department for Environment, Food and Rural Affairs (DEFRA). Agriculture in the English Regions 2011—2nd Estimate. DEFRA, 2013. Available online: https://data.gov.uk/dataset/agriculture_in_the_english_regions (accessed on 11 March 2013).

18. Department for Environment, Food and Rural Affairs (DEFRA). Farm Business Survey for England 2009/10. DEFRA, 2013. Available online: http://www.gov.uk/government/collections/farm-business-survey (accessed on 8 April 2013).

19. Racsko, P.; Szeidl, L.; Semenov, M. A serial approach to local stochastic weather models. *Ecol. Model.* **1991**, *57*, 27–41. [CrossRef]

20. Semenov, M.A.; Barrow, E.M. Use of a stochastic weather generator in the development of climate change scenarios. *Clim. Chang.* **1997**, *35*, 397–414. [CrossRef]

21. Houghton, J.T.; Ding, Y.; Griggs, D.J.; Noguer, M.; van der Linden, P.J.; Dai, X.; Maskell, K.; Johnson, C.A. (Eds.) *Climate Change 2001: The Scientific Basis*; Contribution of Working Group I to the Third Assessment Report of the Intergovernmental Panel on Climate Change; Cambridge University Press: Cambridge, UK, 2001; Available online: http://www.grida.no/publications/other/ipcc_tar/?src=/climate/ipcc_tar/ (accessed on 15 December 2016).

22. Daccache, A.; Sataya, W.; Knox, J.W. Climate change impacts on rainfed and irrigated rice yield in Malawi. *Int. J. Agric. Sustain.* **2015**, *13*, 87–103. [CrossRef]

23. Daccache, A.; Weatherhead, E.K.; Lamaddalena, N. Climate change and the performance of pressurized irrigation water distribution networks under Mediterranean conditions: Impacts and adaptations. *Outlook Agric.* **2010**, *39*, 277–283. [CrossRef]

24. Nakićenović, N.; Davidson, O.; Davis, G.; Grübler, A.; Kram, T.; Lebre La Rovere, E.; Metz, B.; Morita, T.; Pepper, W.; Pitcher, H.; et al. (Eds.) *IPCC Special Report on Emissions Scenarios*; Cambridge University Press: Cambridge, UK, 2000. Available online: http://www.ipcc.ch/ipccreports/sres/emission/index.php?idp=0 (accessed on 15 December 2016).

25. Intergovernmental Panel on Climate Change (IPCC), Task Group on Scenarios for Climate Impact Assessment (TGICA). *Guidelines on the Use of Scenario Data for Climate Impact and Adaptation Assessment*, Version 2; IPCC, 2007. Available online: http://www.ipcc-data.org/guidelines/TGICA_guidance_sdciaa_v2_final.pdf (accessed on 15 December 2016).

26. Pérez-Ortolá, M.; Daccache, A.; Hess, T.M.; Knox, J.W. Simulating impacts of irrigation heterogeneity on onion (Allium cepa L.) yield in a humid climate. *Irrig. Sci.* **2014**, *33*, 1–14.

27. Vanuytrecht, E.; Raes, D.; Willems, P. Global sensitivity analysis output from the water productivity model. *Environ. Modell. Softw.* **2014**, *51*, 323–332. [CrossRef]

28. El Chami, D.; Knox, J.W.; Daccache, A.; Weatherhead, E.K. The economics of irrigating wheat in a humid climate—A study in the East of England. *Agric. Syst.* **2015**, *133*, 97–108. [CrossRef]

29. Mark, T.; Antony, B. Abiotic stress tolerance in grasses from model plants to crop plants. *Plant Physiol.* **2005**, *137*, 791–793.

30. Ashraf, M.Y. Yield and yield components response of wheat (*Triticum aestivum* L.) genotypes tinder different soil water deficit conditions. *Acta Agron. Hung.* **1998**, *46*, 45–51.

31. Bailey, R. *Irrigated Crops and Their Management*; Farming Press: Ipswhich, UK, 1990.

32. Food and Agriculture Organization of the United Nations (FAO). *Sustainable Assessment of Food and Agriculture Systems. SAFA Indicators*; FAO Natural Resources Management and Environment Department, 2012. Available online: http://www.fao.org/fileadmin/templates/nr/sustainability_pathways/docs/SAFA_Guidelines_12_June_2012_final_v2.pdf (accessed on 15 December 2016).

33. Hayati, D.; Ranjbar, Z.; Karami, E. Measuring agricultural sustainability. In *Biodiversity, Biofuels, Agroforestry and Conservation Agriculture*; Lichtfouse, E., Ed.; Sustainable Agriculture Reviews 5; Springer Science & Business Media, 2010; pp. 73–100.

34. Home-Grown Cereal Authority (HGCA). *Market Data Centre: The Information Resource Centre for Cereal and Oilseeds Markets*; HGCA, 2013. Available online: https://cereals.ahdb.org.uk/tools/market-data-centre.aspx (accessed on 8 April 2013).

35. Agro Business Consultants (ABC). *The Agricultural Budgeting & Costing Book*; 75; Agro Business Consultants: Melton Mowbray, UK, 2012.

36. Nix, J. *Farm Management Pocketbook*; 42; Agro Business Consultants: Melton Mowbray, UK, 2011.

37. Environment Agency (EA). *Abstraction Charges Scheme 2013/14*; Environment Agency (EA): Bristol, UK, 2013.

38. Morris, J.; Ahodo, K.; Weatherhead, E.K.; Daccache, A.; Patel A and Knox, J.W. Economics of rainfed and irrigated potato production in a humid environment. In *Economics of Water Management in Agriculture*; Bournaris, T., Berbel, J., Manos, B., Viaggi, D., Eds.; CRC Press: Boca Raton, FL, USA, 2014; pp. 71–97.

39. Ackerman, F.; Stanton, E.A. The social cost of carbon. *Real-World Econ. Rev.* **2010**, *53*, 129–143.

40. Department for Environment, Food and Rural Affairs (DEFRA). Government Conversion Factors for Company Reporting. Greenhouse Gas Conversion Factor Repository. DEFRA, 2016. Available online: http://www.ukconversionfactorscarbonsmart.co.uk (accessed on 14 December 2016).

41. Department of Energy & Climate Change (DECC). A Brief Guide to the Carbon Valuation Methodology for UK Policy Appraisal. DECC, 2011. Available online: http://www.gov.uk/government/collections/carbon-valuation--2 (accessed on 14 December 2016).

42. Hsiao, T.C.; Steduto, P.; Fereres, E. A systematic and quantitative approach to improve water use efficiency in agriculture. *Irrig. Sci.* **2007**, *25*, 209–231. [CrossRef]

43. Sheppard, D.; El Gamal, R. Falling Oil Price Tilts Political, Economic Balance in U.S. Favour. *Reuters News*, 12 September 2014. Available online: http://uk.reuters.com/article/oil-politics-idUKL1N0RC1QE20140912 (accessed on 14 December 2016).

44. Stevens, P. Saudi Arabian Oil Policy: Its Origins, Implementation and Implications. In *State Society and Economy in Saudi Arabia*; Niblock, T., Ed.; Croom Helm: London, UK, 1981.

45. Stevens, P. *Oil and Portics: The Post-War Gulf*; Royal Institute of International Affairs: London, UK, 1992.

46. International Energy Agency (IEA). *World Energy Outlook 2013*; OECD/IEA: Paris, France, 2013; Available online: https://www.iea.org/publications/freepublications/publication/WEO2013.pdf (accessed on 14 December 2016).

47. Amosson, S.H.; Almas, L.; Girase, J.R.; Kenny, N.; Guerrero, B.; Vimlesh, K.; Marek, T. *Economics of Irrigation Systems*; Texas A&M, Agrilife Extension: College Station, TX, USA, 2011.

48. Food and Agriculture Organization of the United Nations (FAO). World Food Situation: FAO Food Price Index. FAO. Available online: http://www.fao.org/worldfoodsituation/foodpriceindex/en/ (accessed on 13 March 2013).

49. Organisation for Economic Co-operation and Development (OECD)/Food and Agriculture Organization of the United Nations (FAO). *Agricultural Outlook 2013–2022*; OECD Publishing: Paris, France, 2013; Available online: http://www.oecd.org/berlin/OECD-FAO%20Highlights_FINAL_with_Covers%20(3).pdf (accessed on 15 December 2016).

50. United States Department of Agriculture (USDA). *USDA Agricultural Projections to 2022*, Long-term Projections Report OCE-2013-1; Office of the Chief Economist, United States Department of Agriculture (USDA): Washington, DC, USA, 2013. Available online: http://www.usda.gov/oce/commodity/projections/USDAAgriculturalProjections2022.pdf (accessed on 15 December 2016).

51. Willenbockel, D. *Exploring Food Price Scenarios Towards 2030 with a Global Multi-Region Model*; Oxfam Research Reports; Oxfam: Oxford, UK, 2011. Available online: https://www.oxfam.org/sites/www.oxfam.org/files/rr-exploring-food-price-scenarios-010611-en.pdf (accessed on 14 December 2016).

52. He, Y.; Hou, L.; Wang, H.; Hu, K.; McConkey, B. A modelling approach to evaluate the long-term effect of soil texture on spring wheat productivity under a rain-fed condition. *Nat.-Sci. Rep.* **2014**, *4*, 5736. [CrossRef] [PubMed]

53. Semenov, M.A.; Shewry, P.R. 2011 Modelling predicts that heat stress, not drought, will increase vulnerability of wheat in Europe. *Nat.-Sci. Rep.* **2011**, *1*, 66.

54. Kersebaum, K.C.; Nendel, C. Site-specific impacts of climate change on wheat production across regions of Germany using different CO_2 response functions. *Eur. J. Agron.* **2014**, *52*, 22–32. [CrossRef]

55. Richter, G.M.; Semenov, M.A. Modelling impacts of climate change on wheat yields in England and Wales: Assessing drought risks. *Agric. Syst.* **2005**, *84*, 77–97. [CrossRef]

56. Dono, G.; Cortignani, R.; Doro, L.; Giraldo, L.; Ledda, L.; Pasqui, M.; Roggero, P.P. Adapting to uncertainty associated with short-term variability changes in irrigated Mediterranean farming systems. *Agric. Syst.* **2013**, *117*, 1–12. [CrossRef]

57. Pachauri, R.K.; Reisinger, A. (Eds.) *IPCC Climate Change 2007: Synthesis Report*; Contribution of Working Groups I, II and III to the Fourth Assessment Report of the Intergovernmental Panel on Climate Change (IPCC); IPCC: Geneva, Switzerland,

2007. Available online: https://www.ipcc.ch/publications_and_data/publications_ipcc_fourth_assessment_report_synthesis_report.htm (accessed on 15 December 2016).

58. Weatherhead, E.K.; Knox, J.W. Predicting and mapping the future demand for irrigation water in England and Wales. *Agric. Water Manag.* **2000**, *43*, 203–218. [CrossRef]

59. Niero, M.; Ingvordsen, C.H.; Peltonen-Sainio, P.; Jalli, M.; Lyngkjær, M.F.; Hauschild, M.Z.; Jørgensen, R.B. Eco-efficient production of spring barley in a changed climate: A life cycle assessment including primary data from future climate scenarios. *Agric. Syst.* **2015**, *136*, 46–60. [CrossRef]

60. Doll, P. Impact of climate change and variability on irrigation requirements: A global perspective. *Clim. Chang.* **2002**, *54*, 269–293. [CrossRef]

Designing a Pattern of Participatory Management with the Approach of Economic Sustainability in Northern Iran (Case Study: Fereydoon Kenar Wetland)

Najmeh Daryaei, Mehdi Mirdamadi, Jamal F. Hosseini, Samad Rahimi Soureh and Reza Arjomandi

Abstract: Wetlands are under increasing pressure from population growth, poverty and economic inequality; social and economic conflicts between local communities; and unsustainable use of plant and animal resources. Agriculture is considered to be one of the important factors affecting wetlands. The overall purpose of this research is to design a participatory management pattern for rice farmers considering the sustainability of Fereydoon Kenar's wetland site in Mazandaran Province, Iran. The entire population are landowners on the wetland site ($N = 3249$), of which 345 rice farmers were selected as a sample using the Cochran formula. To analyze the data from the questionnaires, $SPSS_{Win19}$ and Amos Software were used. Based on the farmers' opinions, development of nursery, employment of rice farmers during the second six month of the year after the rice harvest, and tree planting in the main habitats of the birds (*Damgahs*) in order to exploit their benefits were considered to be the most important indicators of economic sustainability in the Fereydoon Kenar wetland. The results of the path analysis showed that the environmental, structural, educational and policy-making mechanisms had significant effects on the rice farmers' knowledge about participatory management of wetlands. Meanwhile, the structural, educational, and cultural mechanisms, and knowledge about participatory management had significant effects on rice farmers' attitudes towards participatory management in sustainability of the wetland. In addition, the structural mechanisms, knowledge of participatory management in economic sustainability, and attitudes of rice farmers towards participatory sustainable management had a significant and positive effect on the behavior of participatory management.

---

1. Introduction

Wetlands are among the most important ecosystems and life zones in the world, which absolutely have no replacement. The ecological value of wetlands is ten times more than forests and 200 times greater than agricultural lands. Wetlands are one of the most important basic environmental resources. As they are located in the lowest parts of the watersheds, they are usually affected by changes and evolutions of the upstream. This can cause problems such as reduction in water entering the wetlands from surface water resources and groundwater watersheds and plains around the wetlands (failing to meet environmental water rights of wetlands), especially due to dam construction projects and inter-basin water transmission; destruction and sedimentation of wetland area; pollutions caused by urban, industrial and agricultural activities; unsustainable exploitation of plant and animal species in the wetland, particularly illegal and excessive hunting and fishing and harvesting of forage and other wetland products more than wetland's renewable ability; the development of unsustainable activities in wetlands; and the access of non-native and raider species into the wetlands, particularly *Azolla* [1].

The existence of wetland areas in northern Iran and the southern shores of the Caspian Sea receive a significant population of immigrant waterfowl every year and these aquatic habitats form the shelter of diverse plants and animal species. Among the northern Iranian wetlands, Fereydoon Kenar's wetland site in Mazandaran Province, with an area equivalent to 5427 hectares, consists of Fereydoon Kenar, Ezbaran, and Western and Eastern Sorkh Rud Ab-Bandans, which, annually, receive nearly one-third of overwintering bird species (150 species) in Iran and also have the most extensive methods for traditional hunting of birds. This wetland complex comprises different types of artificial or man-made wetlands and is mainly with private ownership, which, in fact, is the same agricultural lands revolving around rice production that are dedicated to the cultivation of this crop during spring and summer and are submerged in autumn and winter by the water of streams and rivers and, based on its breadth and depth, can store different amounts of water (Figure 1). According to the existence of the remains of rice, including seeds or stalks as well as benthic organisms, these lands are considered as a habitat and overwintering place for migratory aquatic birds and other creatures that live beside water. Fereydoon Kenar's wetland site is also considered an international wetland and has a vital importance by having the mentioned conditions, thus discussion of its management, through participation and involvement of stakeholders is essential.

The preservation of natural resources in this wetland, particularly the related plants and animal species including waterfowl, and the dependency of resident's livelihoods on sources of income obtained from hunting the birds of the area, by considering the social conditions and existed beliefs and values in the region, determines the undeniable role of the participation of local communities, especially rice farmers of the region as the owners of wetland areas, more than ever before. Therefore, to protect and preserve the stable survival of this collection depends on the participation and wise use of wetland resources by local people. Several cases are considered in participative management of wetlands that can be classified in form of economic, environmental, structural, educational, policy-making and cultural mechanisms. Careful and detailed examination of each of these mechanisms prepares the ground work for designing an optimal model of participatory management for the rice farmers of this wetland site regarding the wise use of resources and the realization of sustainable development goals related to wetlands in all three—economic, social and environmental—dimensions. The aim of this study is thus the design of a participatory management model of rice farmers for the economic sustainability of Fereydoon Kenar's wetland site in the Mazandaran Province.

(a)   (b)

Figure 1. Fereydoon Kenar wetland before (**a**) and after (**b**) water birds migration.

In his research on the wetlands of the Kisii area in Kenya, Mironga found that limiting public access to and non-normative use of wetland resources reduced wetland destruction to a minimum [2]. Robinson et al. believe that structural or institutional factors such as the lack of appropriate and enforceable property rights, conflicting interests of stakeholders and problems in clear definition of borders affect the development of the management plan in the wetland [3]. Maclean et al., in their research in Uganda, found that ill-defined property rights are often associated with wetland drainage and unsustainable levels of resource use. The property rights issues in Ugandan wetlands are afflicted by contradiction and tensions. For example, there is considerable ambiguity surrounding the concept of government or local government holding wetlands "in trust for the people", and confusion over rights and obligations of ownership on the one hand and management on the other. Substantial new legislation affecting land tenure and use has been set, but still needs to be absorbed by all levels of society [4]. Hannan Khan, in a study examining the participatory management of wetland resources in Bangladesh, found that the establishment of a state multi-stakeholder system as a structure and institutional process for stable wetland resource management is essential. The approach of natural resources management requires the integration of bottom-up and top-down approaches, which must include the interests of all stakeholders in decision-making processes relating to wetlands [5]. Legal arrangements for managing wetlands within nation states are not intrinsically different. The legal structures in each state reflect the historical, cultural, political and constitutional background out of which they have developed. In response to these factors, wetlands are managed in accordance with a matrix of constitutional, strategic, regulatory and management rules. Each of these rules performs a different function within each system. Some are aspirational, while others are informative; some are facilitative, while others are directional; and few are mandatory, while others are cast as enforceable obligations, breach of which leads to a liability in one form or another. Importantly, these rules interact with and inform each other, and yet perform different functions in ways that point in the direction of a relatively coherent system [6]. Abila identifies the lack of clear policies in guiding the use of wetland resources as the main challenge in Kenyan wetlands protection program [7]. Mironga found that the strong and continuous monitoring in exploitation of wetlands by beneficiary groups can prevent imposing pressure on these resources [2]. In addition to coordination between policies implemented regarding wetlands and their management, executive sponsorship for a set of rules in this field are most important policy-making mechanisms [8]. Cultural values can encourage proper behavioral responses relative to the incidence of environmental changes [9]. The Ramsar Convention about the importance of knowledge, belief systems and social practices of local people in the management of wetlands held that local knowledge and skills are made available to assist in the ongoing identification of problems and solutions in the management of wetlands. Often, this information is difficult to access and special participatory processes are needed to bring it to the surface [10]. Table 1 shows a summary of the economic mechanisms extracted from various studies.

Table 1. Research background of economic mechanisms affecting participatory management of farmers in sustainability of wetland sites in the world.

Mechanism	Extracted Item	Researcher(s)
Economic	• Penalties imposed by the government for those who degrade the environment of wetlands	Macharia et al. [11]; Shrestha [12]
	• Provide public funds to develop the site wetland management plans	Lim and McAleer [13]
	• Give financial incentives to universities, academic institutions and research centers based studies in the field of wetland ecology, biodiversity and ecological functions of wetlands, wetland restoration and sustainable utilization of wetlands	Hailun and Dong [14]
	• Economic valuation of the wetland sources	Gawler [15]

Table 1. *Cont.*

Mechanism	Extracted Item	Researcher(s)
Environmental	• Divide the wetland into three functional areas: core (given the sensitive and vulnerable habitat), buffer and development areas	Hailun and Dong [14]
	• Provide sufficient information for farmers about the rules of the wetlands, especially hunting and fishing rules	Darradi et al. [16]
	• Increase awareness of farmers about the methods of improving the management of wetland resources and introduce alternative revenue options to reduce dependence on wetlands	Diouf [17]; Amaniga Ruhanga and Iyango [18]
	• Create awareness of all managers and decision-makers about pollution control, waste management and environmental management in order to prevent unnecessary damage to the wetlands	Kyarisiima et al. [8]
Structural	• Restrict non-normative access and use and prevent illegal use of wetlands resources by individuals and institutions	Turner et al. [19]; Mironga [2]
	• Create adequate management structures with harmonized goals and preferences with the environment in order to manage the wetland site	Robinson et al. [3]; Hannan Khan [5]
	• Clearly define the rights and obligations of land ownership and management of wetland sites	Maclean et al. [4]
	• Work in accordance with local institutions and their role in wetland management	Dixon [20]; Wood et al. [21]; Rahman and Begum [22]
Educational	• Train locals in guiding the tourists to visit the region and aquatic migratory birds	Macharia et al. [11]
	• Learn the skills of the locals in order to expand self-employment	Wood et al. [21]
	• Integrate existing knowledge with the use of various methods of science	Murdiyarso et al. [23]
	• Introduce and include topics related to wetlands and protect them in environmental education topics provided in elementary and high school grades	Maclean et al. [4]

Table 1. *Cont.*

Mechanism	Extracted Item	Researcher(s)
Policy-making	• Periodically monitor and evaluate management plans to be carried out at the level of the wetland	Stanic [24]
	• Define exactly the method of implementing the laws as executive devices, especially strict rules regarding the prohibition of illegal hunting	Abila [7]; Kyarisiima et al. [8]; Fisher [6]; Mapedza et al. [25]
	• Deal with violators of hunting rules with the support of all relevant institutions	Hailun and Dong [14]; Moses [26]
	• Formulate a comprehensive policy for wetlands conservation	Katerere [27]; Dekens et al. [28]
	• Implement a strong and continuous supervision over the operation of the wetland, especially waterfowl hunting and fishing methods	Mironga [2]
	• Equip the local environmental community with information and tools for monitoring the wetland environment	Alberta government [29]
Cultural	• Hold different ceremonies and celebrations in the area	Schuyt and Brander [30]
	• Strengthen the culture of prevention of hunting birds with illegal methods	Casagrande [9]; Shrestha [12]
	• Employ people's religious beliefs about preserving the environment	Schuyt and Brander [30]
	• Review and introduce the history of the wetland sites	

2. Materials and Methods

The present study is considered as applied research in terms of the results, as combinatory research in terms of implementation process, as deductive research in terms of implementation logics, as prospective study in terms of time, and also in terms of the purpose as analytical non-experiential research. Statistical population of this study is composed of all paddy farmers and landowners of Fereydoon Kenar wetland site in Mazandaran Province ($N = 3249$). According to the Cochran formula [31], the sample size calculated for this study equals 345 farmers with whom interviews were conducted. In addition, stratified sampling proportionate to the size of sampling was used to select the samples and the samples were selected using simple random sampling method in each class. Data were collected with a questionnaire, the validity of which was approved by a panel of experts and its reliability was confirmed by carrying out a pilot study and calculating its Cronbach alpha coefficient. The Cronbach alpha values for different parts of the questionnaire exceed 0.7. SPSS$_{\text{Win19}}$ (*IBM SPSS Statistics for Windows*, Version 19.0.; IBM Corp.: Armonk, NY, 2010) and Amos software (Amos, Version 23.0; IBM SPSS; Chicago, IL, USA, 2014) have been used for data analysis.

3. Results

The results of the farmers' analysis by age showed that their average age was over 52 years. Maximum age was 75 years and minimum age was 31 years. In addition, the majority of farmers had certificate of junior high school or lower levels of education. On average, their families consisted of four members (Table 2).

Table 2. Personal and professional characteristics of rice farmers.

Variable	Group	Frequency	Percent	Mean	SD	Min	Max
Age (year)	<40	59	17.1				
	41–50	122	35.4				
	51–60	84	24.3	52.130	10.606	31	75
	61–70	63	18.3				
	>70	17	4.9				
Education Level	Illiterate	50	14.5				
	Primary school	106	30.7				
	Junior high school	87	25.2	—	—	—	—
	High school	81	23.5				
	High school diploma and upper	21	6.1				
Number of Household Members	2–3	107	31				
	4–5	172	49.9	—	—	2	8
	6–7	63	18.3				
	8	3	0.9				
Experience of Rice Farming (year)	<20	102	29.6				
	20–30	117	33.9				
	30–40	51	14.8	29.229	13.251	7	62
	40–50	60	17.4				
	50–60	9	2.6				
	>60	6	1.7				
Total Area of Rice Lands (ha)	<1	119	34.5				
	1–2	147	42.6				
	2–3	61	17.7	1.488	0.899	0.4	5
	3–4	10	2.9				
	4–5	8	2.3				

In addition, findings on the evaluation of participatory management of paddy farmers in the economic sustainability of the wetlands using ISDM (Interval of Standard Deviation from the Mean) method showed that 14.2% of respondents had a low level of participatory management behavior, 40.6% had moderately low level of participatory management behavior, 30.7% had moderately high level of participatory management behavior and 14.5% had a high level of participatory management behavior (Table 3).

Table 3. Distribution of the respondents according to the level of participatory management behavior in the economic sustainability of wetlands.

Levels of Behavior	Frequency	Percentage	Cumulative Percentage
Low	49	14.2	14.2
Moderately low	140	40.6	54.8
Moderately high	106	30.7	85.5
High	50	14.5	100
Total	345	100	—

Mean: 31.63; Standard Deviation: 5.43.

In addition, as can be seen in Table 4, the item of nursery development to increase income was specified as the highest priority and the item of mushroom production was specified as the lowest priority in the collection of items related to participatory management behavior of paddy farmers in the economic sustainability of wetlands.

Table 4. Prioritization of items of participatory management behavior of paddy farmers in the economic sustainability of wetland.

Item	Average* (of 5)	Standard Deviation	Coefficient of Variations	Priority
Development of nursery	2.568	0.752	0.292	1
Employment of rice farmers in the second six month of the year after the rice harvest	3.031	0.941	0.310	2
Tree planting in the main habitats of birds (*Damgahs*) in order to exploit its benefits	2.660	0.871	0.327	3
Insurance of land income for all owners of the wetland as agriculture land supplying household income	2.756	0.904	0.328	4
Reconstruction of irrigation projects in the region	2.846	0.965	0.339	5
Development and prosperity of the local markets, especially fish markets	2.863	0.986	0.344	6
Introduction of rice produced in *Damgahs* (trapping site) of Fereydoon Kenar wetland site as organic product with affordable price	2.631	0.931	0.353	7
Local poultry breeding	2.484	0.895	0.360	8
The production of Compost from *Azolla* ferns	2.188	0.870	0.397	9

Table 4. *Cont.*

Item	Average* (of 5)	Standard Deviation	Coefficient of Variations	Priority
Preparing household food by local residents, local and traditional cuisine tour and the establishment of a family hotel in order to increase income	1.944	0.799	0.411	10
Development of natural tourism activities (ecotourism) in the region (bird watching, hiking, fishing and boating, etc.)	1.634	0.686	0.419	11
Constructing tourist huts	2.005	0.889	0.443	12
Mushroom production	2.014	0.965	0.479	13

* Likert range: 1, Very Low; 2, Low; 3, Medium; 4, High; 5, Very High.

In addition, in order to gain a suitable model of the participatory management of paddy farmers in the economic sustainability of Fereydoon Kenar wetland site, and in accordance with the literature and theoretical framework of the research, we tried to select effective variables. Nine variables,

1. participatory management behavior of paddy farmers regarding economic wetland sustainability,
2. paddy farmers' knowledge of participatory management of economic wetland sustainability,
3. paddy farmers' attitudes toward participatory management of wetland sustainability,
4. economic mechanisms,
5. environmental mechanisms,
6. structural mechanisms,
7. educational mechanisms,
8. cultural mechanisms, and
9. policy-making mechanisms

were entered in the model.

Figure 2 shows an analysis diagram of the research path. The indices of model fit and results of the path analysis are presented in Tables 5 and 6.

Table 5. Fit Indices of the Fitted Model.

Index	Optimal Value	Reported Value
X^2/df	<3	1.448
RMSEA	<0.07	0.038
CFI	>0.95	0.999
NNFI or TLI	>0.96	0.993
IFI	0.95	0.999

RMSEA = Root Mean Square Error of Approximation; CFI = Corrected Comparative Fit Index; NNFI or TLI = Non-Normed Fit Index or Tucker Lewis Index; IFI = Incremental Fit Index.

As can be seen in Table 5, proportionate amounts of fit indices include chi-square on the degrees of freedom, RMSEA (Root Mean Square Error of Approximation), CFI (Corrected Comparative Fit Index), NNFI or TLI (Non-Normed Fit Index or Tucker Lewis Index) and IFI (Incremental Fit Index) represent a reasonable adjustment of data-model.

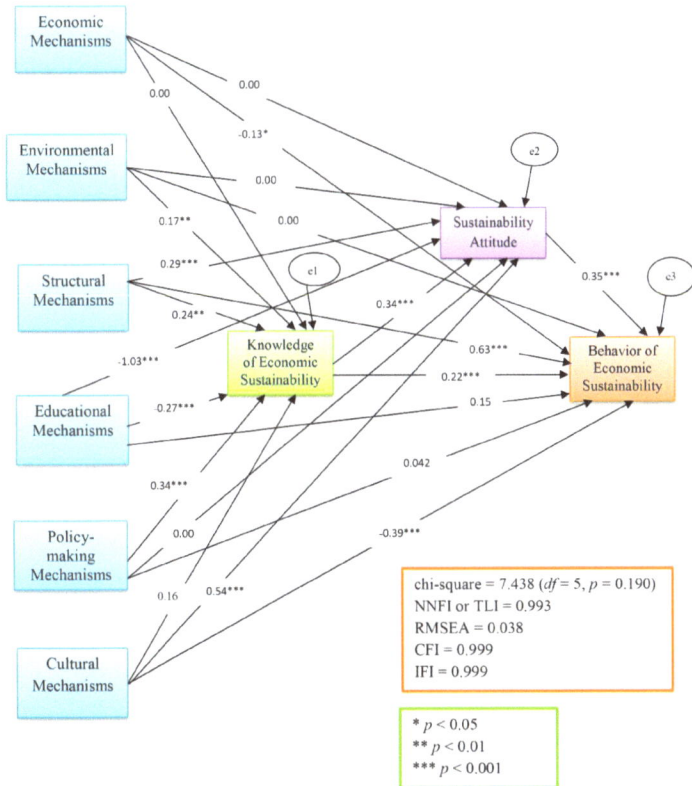

Figure 2. The model of mechanisms affecting participatory management behavior of paddy farmers regarding economic sustainability of wetland. RMSEA = Root Mean Square Error of Approximation; CFI = Corrected Comparative Fit Index; NNFI or TLI = Non-Normed Fit Index or Tucker Lewis Index; IFI = Incremental Fit Index.

Table 6. Analysis of the direct, indirect and total effects in the model of mechanisms affecting participatory management behavior of paddy farmers regarding economic sustainability of wetland.

Variables	Economic Mechanisms	Environmental Mechanisms	Structural Mechanisms	Educational Mechanisms	Policy-making Mechanisms	Cultural Mechanisms	Participatory Management Knowledge	Attitudes towards Participatory Management	Participatory Management Behavior
Direct effects									
Participatory Management Knowledge	0.000	0.174	0.241	−0.271	0.345	0.160	—	—	—
Attitudes towards Participatory Management	0.000	0.000	0.296	−1.034	0.000	0.543	0.349	—	—
Participatory Management Behavior	−0.130	0.000	0.638	0.149	0.042	−0.392	0.221	0.355	-
Indirect effects									
Participatory Management Knowledge	—	—	—	—	—	—	—	—	—
Attitudes towards Participatory Management	0.000	0.060	0.084	−0.094	0.120	0.055	—	—	—
Participatory Management Behavior	0.000	0.059	0.187	−0.459	0.003	0.246	0.123	—	—
Total effects									
Participatory Management Knowledge	0.000	0.174	0.241	−0.271	0.345	0.160	—	—	—
Attitudes towards Participatory Management	0.000	0.060	0.380	−1.128	0.120	0.598	0.349	—	—
Participatory Management Behavior	−0.130	0.059	0.825	−0.310	0.045	−0.146	0.344	0.355	—

In addition, as can be seen in Table 6 and Figure 2, the variable of economic mechanisms had a direct, negative and significant impact on the variable of participatory management behavior of paddy farmers regarding economic wetland sustainability ($\beta = -0.130$, $p < 0.05$) and had no effect on the two variables of paddy farmers' knowledge of participatory management of economic wetland sustainability and paddy farmers' attitudes toward participatory management of wetland sustainability. The variable of environmental mechanisms had a direct, positive and significant impact on the variable paddy farmers' knowledge of participatory management of economic wetland sustainability ($\beta = 0.174$, $p < 0.01$) and had no effect on the variables paddy farmers' attitudes toward participatory management of wetland sustainability and participatory management behavior of paddy farmers regarding economic wetland sustainability. In addition, structural mechanisms had a direct, positive and significant effect on three variables of paddy farmers' knowledge of participatory management of economic wetland sustainability ($\beta = 0.241$, $p < 0.01$), paddy farmers' attitudes toward participatory management of wetland sustainability ($\beta = 0.296$, $p < 0.001$), and participatory management behavior of paddy farmers regarding economic wetland sustainability ($\beta = 0.638$, $p < 0.001$). In other words, this variable can be a good predictor of three variables of paddy farmers' knowledge of participatory management of economic wetland sustainability, paddy farmers' attitudes toward participatory management of wetland sustainability and their participatory management behavior of paddy farmers regarding economic wetland sustainability. Educational mechanisms also had a direct, negative and significant effect on the variables of paddy farmers' knowledge of participatory management of economic wetland sustainability ($\beta = -0.271$, $p < 0.001$) and a direct, negative and significant effect on the paddy farmers' attitudes toward participatory management of wetland sustainability ($\beta = -1.034$, $p < 0.001$). In addition, this variable had no effect on the variable of participatory management behavior of paddy farmers regarding economic wetland sustainability. Policy-making mechanisms also had a direct, positive and significant effect on the variable of paddy farmers' knowledge of participatory management of economic wetland sustainability ($\beta = 0.345$, $p < 0.001$) and no significant effect on the variable of participatory management behavior of paddy farmers regarding economic wetland sustainability. In addition, the variable of policy-making mechanisms had no effect on the variable of paddy farmers' attitudes toward participatory management of wetland sustainability. Cultural mechanisms had a direct, positive and significant effect on the variable of paddy farmers' attitudes toward participatory management of wetland sustainability ($\beta = 0.543$, $p < 0.001$); a direct, negative and significant effect on the variable of participatory management behavior of paddy farmers regarding economic wetland sustainability ($\beta = -0.392$, $p < 0.001$); and no effect on paddy farmers' knowledge of participatory management of economic wetland sustainability.

The variable of paddy farmers' knowledge of participatory management of economic wetland sustainability had a direct, positive and significant effect on the variables of paddy farmers' attitudes toward participatory management of wetland sustainability ($\beta = 0.349$, $p < 0.001$) and participatory management behavior of paddy farmers regarding economic wetland sustainability ($\beta = 0.221$, $p < 0.001$). The variable paddy farmers' attitudes toward participatory management of wetland sustainability had a direct, positive and significant effect on the variable participatory management behavior of paddy farmers regarding economic wetland sustainability ($\beta = 0.355$, $p < 0.001$). In addition, as can be seen in Table 6, policy-making mechanisms had the greatest total effect on the variable of paddy farmers' knowledge of participatory management of economic wetland sustainability. Cultural mechanisms had the greatest total effect (positive) on the variable of paddy farmers' attitudes toward participatory management of wetland sustainability, and the variable of structural mechanisms had the greatest total effect on the variable of participatory management behavior of paddy farmers regarding economic wetland sustainability.

The stepwise method of multivariate regression test was used to determine the ability of the independent variables to predict the variable of participatory management behavior of paddy farmers regarding economic wetland sustainability. As Table 7 shows, the four variables of structural mechanisms, paddy farmers' attitudes toward participatory management of wetland sustainability, and paddy farmers' knowledge of participatory management of economic wetland sustainability were entered into the regression equation. According to the adjusted value of R^2, these variables are able to explain approximately 50% of changes of the variable of participatory management behavior of paddy farmers regarding economic wetland sustainability.

Table 7. The results of the stepwise multiple regression analysis with the dependent variable of participatory management behavior of paddy farmers regarding economic sustainability of wetland.

Independent Variables	B	SE B	Beta	T	Tsig	R	R^2	R^2Adj
Structural Mechanisms (X_1)	0.763	0.078	0.658	9.810	0.002	0.554	0.307	0.305
Attitude towards Participatory Management in Sustainability of Wetland (X_2)	0.266	0.038	0.292	6.941	0.000	0.659	0.434	0.431
Cultural Mechanisms (X_3)	−0.404	0.076	−0.349	−5.303	0.000	0.682	0.465	0.460
Participatory Management Knowledge in the Economic Sustainability of Wetland (X_4)	0.195	0.042	0.234	4.613	0.000	0.705	0.497	0.491
Constant	−7.822	2.542	—	−3.077	0.000	—	—	—

F Value = 83.877, Significance = 0.000.

According to the regression coefficients (B) and calculated constant coefficients, the regression equation is as follows:

$$Y = -7.822 + 0.763X_1 + 0.266X_2 - 0.404X_3 + 0.195X_4 + error$$

4. Discussion and Conclusions

Iran is a country that located at the center of many bird migration paths. In fact, this country is the main corridor of annual birds migration from northern China, Kazakhstan, Turkmenistan, Azerbaijan, Armenia, East and West Siberia to northern Europe, Scandinavian countries located in North-West Europe and Africa. Accordingly, many wetlands of Iran are a safe haven for many migratory birds. Fereydoon Kenar, Ezbaran and Sorkh Ruds Ab-Bandans have become as one of the most important overwintering habitats for migratory birds due to favorable conditions and abundant food resources. Whooper swan, mute swan, Bewick's swan, red-breasted goose, lesser white fronted goose, bean goose, marbled duck, ferruginous duck, white-headed duck, Eurasian bittern, ruddy shelduck and dalmatian pelican are among most important bird species in this region. This wetland is overwintering habitat for the sole survivor of the West Siberian cranes in the world. Conservation of natural resources of this wetland, particularly its plants and animals species including aquatic birds, and its residents' dependence on bird hunting as an income source, considering the social conditions, beliefs and values in the region, further indicate the undeniable role of local communities, especially farmers of the region.. Therefore, protection and sustainable conservation of this wetland site depend on the participation and wise use of the wetland resources by locals.

To achieve the farmers' participatory management pattern in the economical sustainability of Fereydoon Kenar wetland site, path analysis was used. The variables used include the paddy farmers' knowledge of participatory management of economic wetland sustainability; paddy farmers' attitudes toward participatory management of wetland sustainability; economic, environmental, structural, educational, and cultural mechanisms; policy-making; and, the main variable, participatory management behavior of paddy farmers regarding economic wetland sustainability. The results showed that there is a direct negative and significant relationship between the variable of economic mechanisms and the participatory management behavior of paddy farmers regarding economic wetland sustainability. Perhaps one of the main reasons for this relationship is that the farmers focus on their livelihoods too much and they have a great attention and desire to earn short-term gains instead of long-term benefits by the exploitation of wetland resources, and this is practically in conflict with the principles of wetland's economic sustainability. These results are consistent with the findings of Mironga [2] and Kyarisiima et al. [8].

There is a direct positive and significant relationship between the variable of structural mechanisms and the participatory management behavior of paddy farmers regarding economic wetland sustainability. These results are consistent with the findings of Mironga [2] in Western Kenya, Dixon [20] and Wood [21] in Ethiopia, Robinson et al. [3], Shrestha [12] and Hannan Khan [5] in Bangladesh. In other words, structural mechanisms such as restricting public and unprincipled access to wetland resources, paying attention to local institutions in the process of wetland management as a missing link in the development of the relationship between people and the environment, clarity in defining the rights and obligations of land ownership and management of wetland site, and paying attention to the discussion of the use of wetland lands and correcting it, are necessary. Also, establishing a state system as an institutional structure and process in order to stabilize the decisions in the management of wetland resources has positive effect on the farmers' participatory management in the economic sustainability of wetland.

Moreover, there is a direct positive and significant relationship between the variables of paddy farmers' knowledge of participatory management of economic wetland sustainability and their attitudes toward participatory management of wetland sustainability, and participatory management behavior of paddy farmers regarding economic wetland sustainability. In other words, enhancing the paddy farmers' knowledge of participatory management of economic wetland sustainability and their attitudes toward participatory management of wetland sustainability has a positive effect on theoir behavior regarding participatory management in economic wetland sustainabilit. These results are consistent with the findings of Gawler [13], Wood et al. [21], Mironga [2], Shtestha [12] and Murdiyarso et al. [23].

Furthermore, the stepwise multivariate regression test was used to determine the ability of independent variables to predict the dependent variable of participatory management behavior of paddy farmers regarding economic wetland sustainability. The result of this test showed that, according to the adjusted R^2, in total, four variables, structural mechanisms paddy farmers' attitudes toward participatory management of wetland sustainability, cultural mechanisms and paddy farmers' knowledge of participatory management of economic wetland sustainability, can predict approximately 50% of the changes in participatory management behavior of paddy farmers regarding economic wetland sustainability.

5. Recommendations

- Given the direct, positive and significant effect of structural mechanisms of paddy farmers' knowledge, attitudes and behavior regarding participatory management of Fereydoon Kenar's economic wetland sustainability, we suggest the integration of collaborative activities related to wetland management (such as technical and civil projects, holding training courses, etc.), public participation in this field by establishing a distinct and separate governmental structures for wetland management, and creating local and participatory associations. This could be a very good incentive for farmers to participate more in the sustainable management of wetland sites.

- According to the positive and significant effect of the variables paddy farmers' knowledge of participatory management of economic wetland sustainability and the paddy farmers' attitudes toward participatory management of wetland sustainability on participatory management behavior of paddy farmers regarding economic wetland sustainability, we suggest different courses and workshops on new and alternative options for the livelihood of local communities with a focus on the exchange of views among experts, farmers and local menagerie-owners to improve living conditions and increase the income level of residents. This will aid in the preservation and wise use of wetland resources and the promotion of sustainable development of Fereydoon Kenar.
- Considering the lack of studies on the economic and environmental value of Fereydoon Kenar wetland site, we propose separate evaluation of the wetland resources from economic and conservational aspects as the basis for any management action taken in this site.

Author Contributions: The experiment was designed by Najmeh Daryaei as well as Mehdi Mirdamadi and Seyed Jamal Hosseini as main advisors to Najmeh Daryaei. Samad Rahimi Soureh and Reza Arjomandi contributed in writing the paper and the data was analyzed by Najmeh Daryaei and confirmed a by panel of experts. All authors have read and approved the final manuscript.

Conflicts of Interest: The authors declare no conflict of interest.

References

1. Bagherzadeh Karimi, M. Factors of unsustainability in Iran's wetlands, 2012. Available online: http://iranwetland.blogfa.com (accessed on 10 January 2016).
2. Mironga, J.M. Effects of farming practices on wetlands of Kisii district, Kenya. *J. Appl. Ecol. Environ. Res.* **2005**, *3*, 81–91. [CrossRef]
3. Robinson, J.; Dent, J.; Schaffer, G. Integrating scientific assessment of wetland areas and economic evaluation tools to develop an evaluation framework to advise wetland management. School of Economics Discussion Paper No. 420. School of Economics, The University of Queensland: Australia, 2010. Available online: http://www.uq.edu.au/economics/abstract/420.pdf (accessed on 12 February 2016).
4. Maclean, I.M.D.; Tinch, R.; Hassall, M.; Boar, R. *Social and economic use of wetland resources: A case study from lake Bunyonyi, Uganda*; University of East Anglia, Centre for Social and Economic Research on the Global Environment: Norwich, UK, 2003; Available online: http://www.cserge.ac.uk/sites/default/files/ecm_2003_09.pdf (accessed on 10 January 2016).

5. Hannan Khan, S.M. Participatory Wetland Resource Governance in Bangladesh: An Analysis of Community-Based Experiments in Hakaluki Haor. Ph.D. Thesis, University of Manitoba, Winnipeg, MB, Canada, 2011.

6. Fisher, D.E. Managing wetlands sustainably as ecosystems: the contribution of the law (part 2). *J. Water Law* **2010**, *21*, 53–65.

7. Abila, R. *Biodiversity and Sustainable Management of a Tropical Wetland Lake Ecosystem: A Case Study of Lake Kanyaboli, Kenya (Report)*; Department of Zoology, Maseno University: Maseno, Kenya, 2005; Available online: https://www.uni-siegen.de/zew/publikationen/volume0305/abila.pdf (accessed on 8 February 2016).

8. Kyarisiima, C.C.; Nalukenge, I.; Karaiuki, W.; Mesaki, S. Factors Affecting Sustainability of Wetland Agriculture within Lake Victoria Basin in Uganda. *J. Agric. Soc. Res.* **2008**, *8*, 78–88. [CrossRef]

9. Casagrande, D.V. The Human Component of Urban Wetland Restoration. In *Restoration of an urban salt marsh: An interdisciplinary approach*; Bulletin No. 100; Casagrande, D.V., Ed.; Yale School of Forestry and Environmental Studies, Yale University: New Haven, CT, USA, 1997; pp. 254–270.

10. Ramsar Convention Secretariat. *Participatory Skills: Establishing and strengthening local communities' and indigenous people's participation in the management of wetlands, Ramsar Handbooks for the Use of Wetlands*, 4th ed.; Ramsar Convention Secretariat: Gland, Switzerland, 2010; Volume 7.

11. Macharia, J.M.; Thenya, T.; Ndiritu, G.G. Management of highland wetlands in central Kenya: The importance of community education, awareness and eco-tourism in biodiversity conservation. *J. Biodivers.* **2010**, *11*, 85–90. [CrossRef]

12. Shrestha, U. Community participation in wetland conservation in Nepal. *J. Agric. Environ.* **2011**, *12*, 140–147. [CrossRef]

13. Lim, C.; McAleer, M. Use of wetlands for sustainable tourism management (Report). University of Western Australia, 2007. Available online: http://www.iemss.org/iemss2002/proceedings/pdf/volume%20due/391.pdf (accessed on 8 February 2016).

14. Hailun, W.; Dong, X. Construction of Wetland Ecotourism Management System—Case Study for Wetland in Jinyin Lake, Wuhan. In Proceedings of the International Conference on E-Business and E-Government (ICEE), Shanghai, China, 6–8 May 2005; pp. 189–192.

15. Gawler, M. What Are Best Practices? Lessons in Participatory Management of Inland and Coastal Wetlands. In Proceedings of the Workshop Held at the 2nd International Conference on Wetlands and Development, Dakar, Senegal, 10–14 November 1998; Gawler, M., Ed.; The World Conservation Union, Wetlands International, World Wide Fund for Nature (WWF): Wageningen, The Netherlands, 2002; pp. 1–12. Available online: https://portals.iucn.org/library/efiles/documents/2002-012.pdf (accessed on 20 February 2016).

16. Darradi, Y.; Grelot, F.; Morardet, S. Analyzing stakeholders for sustainable wetland management in the Limpopo river basin: The case of Ga-Mampa wetland, South Africa. In Proceedings of the 7th Symposium Mainstreaming IWRM in the Development Process, Lilongwe, Malawi, 1–3 November 2006.

17.   Diouf, A. M. Djoudj National Park and its Periphery: An Experiment in Wetland Co-management. In Proceedings of the Workshop Held at the 2nd International Conference on Wetlands and Development, Dakar, Senegal, 10–14 November 1998; Gawler, M., Ed.; The World Conservation Union, Wetlands International, World Wide Fund for Nature (WWF): Wageningen, The Netherlands, 2002; pp. 13–18. Available online: https://portals.iucn.org/library/efiles/documents/2002-012.pdf (accessed on 20 February 2016).

18.   Amaniga Ruhanga, I.; Iyango, L. A socio-economic baseline survey of communities adjacent to lake Bisina/Opeta and lake Mburo/Nakivali wetland systems. In *Providing Baseline Information for the Implementation of the COBWEB Project in Western and Eastern/North-Eastern Uganda*; The East Africa Natural History Society: Kampala, Uganda, 2010.

19.   Turner, R.K.; Van den berg, J.C.; Soderqvist, T.; Barendregt, A.; van der straiten, J.; Maltby, E.; van ierland, E.C. Ecological-economic analysis of wetlands: scientific integration for management and policy. *J. Ecol. Econ.* **2000**, *35*, 7–23. [CrossRef]

20.   Dixon, A.B.; Wood, A.P. Local institutions for wetland management in Ethiopia: Sustainability and state intervention. In *Community-Based Water Law and Water Resource Management Reform in Developing Countries*; Wallingford Publisher: Wallingford, UK, 2007; pp. 130–145.

21.   Wood, A.; Hailu, A.; Abbot, P.; Dixon, A. Sustainable management of wetlands in Ethiopia: Local knowledge versus government policy. In Proceedings of the Workshop Held at the 2nd International Conference on Wetlands and Development, Dakar, Senegal, 10–14 November 1998; Gawler, M., Ed.; The World Conservation Union, Wetlands International, World Wide Fund for Nature (WWF): Wageningen, The Netherlands, 2002; pp. 81–88. Available online: https://portals.iucn.org/library/efiles/documents/2002-012.pdf (accessed on 20 February 2016).

22.   Rahman, M.M.; Begum, A. The strategy of empowering poor for wetland resources conservation in Bangladesh. *J. Hum. Ecol.* **2010**, *31*, 87–92.

23.   Murdiyarso, D.; Kauffman, J.B.; Warren, M.; Pramova, E.; Hergoualc'h, K. Tropical wetlands for climate change adaptation and mitigation: Science and policy imperatives with special reference to Indonesia (CIFOR Working Paper). Center for International Forestry Research: Bogor, Indonesia, 2012. Available online: http://www.cifor.org/publications/pdf_files/WPapers/WP91Murdiyarso.pdf (accessed on 10 January 2016).

24.   Stanic, M. Opportunities and challenges of tourism in wetland areas (Report), University of Aveiro, Portugal, 2013. Available online: https://sustainabledevelopment.un.org/content/documents/1792Ramsar_UNWTO_tourism_E_Sept2012.pdf (accessed on 25 January 2016).

25.   Mapedza, E.; Chisaka, J.; Koppen, B.V. Competing livelihood strategies in the Lukanga wetlands: Reflections from Kapukupuku and Waya area of Zambia. In Proceedings of the 8th WaterNet/Warfsa/GWP-SA Symposium, Lusaka, Zambia, 31 October–2 November 2007.

26. Moses, O. An institutional analysis of the management of wetland resources: A comparative study of Floahreppur municipality in south Iceland and Oyam district in Uganda, Land Restoration Training Program Keldnaholt (Final Project), 2008. Available online: http://www.unulrt.is/static/fellows/document/moses.pdf (accessed on 8 February 2016).

27. Katerere, J.M. Participatory natural resource management in the communal lands of Zimbabwe: What role for customary law? *Afr. Stud. Q.* **2001**, *5*, 115–138.

28. Dekens, J.; Nazoumou, Y.; Zamudio, N.; Adamou, M.M.; Hambally, Y.; McCandless, M. Sustainable wetland management in the face of climate risks in Niger: The case of La Mare De Tabalak, 2013. International Institute for Sustainable Development (IISD). Available online: http://www.iisd.org/library/sustainable-wetland-management-face-climate-risks-niger-case-la-mare-de-tabalak (accessed on 15 February 2016).

29. Alberta Government (2013) Alberta wetland policy (Report 2013), Canada. Available online: http://aep.alberta.ca/water/programs-and-services/wetlands/documents/AlbertaWetlandPolicy-Sep2013.pdf (accessed on 8 February 2016).

30. Schuyt, K.; Brander, L. *The Economic Values of the World's Wetlands*; World Wildlife Fund (WWF): Gland, Switzerland, 2004.

31. Cochran, W.G. *Sampling Techniques*; Harvard University: Cambridge, MA, USA, 1977.

The Use of Indigenous Knowledge in Subsistence Farming: Implications for Sustainable Agricultural Production in Dikgale Community in Limpopo Province, South Africa

Sejabaledi Agnes Rankoana

Abstract: The present study examined community-based mechanisms of continued subsistence farming under unfavourable environmental conditions. Semi-structured interviews conducted with a sample of 250 participants showed that community members sustain farming through their indigenous knowledge. Community members continue subsistence farming in their home-gardens and ploughing fields through indigenous farming practices and rainfall prediction. The practices involve improvement of soil fertility and structure, maintenance of crops, and seed selection and storage for future planting. Knowledge of rainfall prediction is helpful in planning the planting season. These indigenous practices could be helpful in the achievement of the United Nations' Sustainable Development Goal on food security, which requires a nutritionally adequate and safe food supply at household levels.

1. Introduction

Indigenous knowledge is a systematic body of knowledge acquired by local people through accumulation of experience, informal experiment, and understanding of their environment [1]. The indigenous systems of crop production emerged over centuries of cultural and biological evolution and represent the accumulated experiences of indigenous farmers. The farmers produce indigenous crops through knowledge of environmental conditions and seasonal change without access to external inputs, capital, and modern scientific knowledge [2]. After centuries of cultural and biological evolution, communities have developed locally-adapted, complex farming systems that have helped them manage a variety of environments to meet their subsistence needs [3]. Indigenous crop production provides rural communities with food resources [4].

According to Azam-Ali [5], production of indigenous crops forms the basis of subsistence agriculture in sub-Saharan Africa. Food is produced from cereals, legumes, cucurbits, cowpeas and groundnuts, bulrush millet, finger millet, sorghum, gourds, melons, and pumpkins. These crops are produced in subsistence farms and home-gardens as abundant food sources. For Sigot [6], throughout the African continent, small plots of land near homesteads are used as home-gardens. The gardens provide households with a range of plants that provide food. Van Wyk [7] posits that indigenous cereals provide food security to small-scale farmers because they are tolerant of poor soil and drought.

Subsistence farming is rapidly disappearing due to major social, political, and economic changes [8]. Conservation and management of subsistence farming practices may be possible only if they are linked to the preservation of the cultural diversity and economic viability of the local farming populations [8]. Rook [9] observes that many indigenous crop production systems are characterized by low productivity and instability of production. Marginal and erratic rainfall is responsible for poor crop productivity. Poor growth and yield are attributed to low soil and ambient temperatures which drop below the minimum root and shoot growth temperature of 10 °C [10,11]. The inadequacy and uncertainty of rainfall and its uneven and irregular distribution is compounded by low fertility and high fragility of soils.

However, indigenous knowledge enables its owners to enhance subsistence farming at the time of seasonal and climatic variability [12]. Subsistence farming is sustained through indigenous adaptation mechanisms [13,14], which Reid and Huq refer to as "community-based adaptation" that "can be defined as 'community-led process, based on communities' priorities, needs, knowledge, and capacities, which should empower people to plan and cope with the impacts of climate change'" [15]. Indigenous knowledge used in subsistence farming could be promoted as an adaptive mechanism to sustain the livelihood of rural communities with the potential of securing food [16]. The present study examined community-based mechanisms of continuing subsistence farming under unfavourable environmental conditions.

2. Methods

2.1. Study Area

The study was conducted among the Northern Sotho of Dikgale community in Capricorn District of the Limpopo Province, South Africa. Dikgale community is located within Polokwane Local Municipality, approximately 40 km from Polokwane City, and 15 km from the University of Limpopo in Mankweng Township. The community covers an area of 71 km² (6 km × 10.8 km). It is situated between 23.46°–23.48° south latitude and 29.42°–29.47° east longitude. It lies at an average altitude of 1400 above the mean sea level. The study area is on the Highveld Plateau, which is bounded in the south and southeast by the Strydpoort Mountains and in the east and northeast by the Wolkberge. Dikgale area has an annual rainfall of approximately 505 mm. It has a daily average summer temperature of between 16 °C and 27 °C, with the average winter temperature between 5 °C and 19 °C. Summer rainfall occurs between October and April, followed by a dry winter season [17].

Dikgale community has a population of about 45,083 with a population density of 116 per km^2. The primary language spoken by community members is Sepedi. Dwelling units consist of conventional brick houses and fewer huts. The residential area is made up of demarcated housing stands with a block of ploughing fields in a flatter and sandy area [18]. A small number of community members grow subsistence crops in the home-gardens and fields to provide the additional dietary requirements of a balanced intake [18,19].

2.2. Study Design

A qualitative study was conducted to examine sustainable production of subsistence crops. Data were collected through direct interactions with participants.

2.3. Participants

The study population was Dikgale community. A short survey was conducted in the community prior to data collection with the objective of identifying potential units of analysis. The potential units of analysis were residents in Dikgale community, whose home-gardens and fields have signs of recent subsistence crop production. Two hundred and fifty households were purposely selected. Semi-structured interviews were conducted with 250 bread-winners drawn from the households. The participants aged between 25 and 87 years. They were 98 males and 152 females. They consented to participate in the study via signing of the standard university consent form. The interviews were conducted in Sepedi, the local dialect; in the comfort of the participants' own households.

2.4. Data Collection

Semi-structured interviews were used to collect data. Three master's students were trained and have assisted with data collection and analysis. The participants were asked questions about the indigenous knowledge used to endure subsistence farming. Data analysis was effected through a computerized software package and content analysis.

2.5. Quality Criteria

Comprehensiveness and trustworthiness of collected data were attained through reviews of data with the participants. The participants were able to provide corrections to the inconsistencies, contradictions, and data gaps. Consultations were made to validate and clarify data.

3. Results and Discussion

3.1. Subsistence Farming

The participants were asked to describe subsistence farming. The responses to this question reflect that community members provide supplemental food for their families through production of indigenous crops in the home-gardens surrounding the compound and ploughing fields allocated to each household by the chief-in-council. The fields are usually located within reasonable walking distance from the villages and are arranged in a rectangular pattern. An average land holding is two hectares per household. Strips of uncultivated grassland separate the fields. Tractors are available for hire at a cost of about 900 South African Rand (ZAR) per hectare. Planting of crops generally commences after the first rain has fallen. The seeds are mixed and sown simultaneously to grow the crops together in the same field.

3.2. The Use of Indigenous Knowledge to Sustain Subsistence Farming

The participants were asked questions relating to the indigenous knowledge community members use to sustain subsistence farming. The responses provided show that community members use their indigenous farming practices such as planting on different soil types, soil fertilization, selection and storage of seeds and maintenance of crops. In addition to these knowledge systems the participants mentioned the use of knowledge of rainfall forecast. These indigenous knowledge systems are produced by local people based on their lived experiences [20]. The Food and Agricultural Organisation (FAO) [21] attests that local farmers and indigenous communities have indigenous knowledge, expertise, skills, and practices related to sustainable agricultural production.

3.2.1. Rainfall Forecast

It was established during the study that the participants use signs, such as the sprouting tree leaves and flowers, to predict rainfall. The beginning of summer is marked by the flowing and leaf sprouting of *Senegali* species. In addition to this plant phenology, the appearance of stars, moon, and the sun are carefully observed at the beginning of September, which marks the beginning of a new season. It is believed that the signs of these celestial bodies denote a good or bad season. For example, 78% of the participants showed that if mahlapolane (Mars) lies towards the west, it predicts a good year, but if it disappears towards the east, it predicts a bad season with little rain. It is also believed that if the horns of the crescent moon point towards the Earth, it pours out rain, but if it points away from the Earth, it holds the rain. This indigenous knowledge of forecasting is used to plan the planting season. Whenever a bad season is predicted, farming will not be done until it rains sufficiently.

The participants' knowledge of rainfall prediction corroborates Speranza et al.'s [22] findings that local farmers possess knowledge on the use of local indicators, such as plants, birds, insects, and astronomy, in predicting rainfall. Kijazi et al. [23] attest that people use the behaviour of animals and plants to predict the coming agricultural season. Chang'a et al. [24] show that this type of indigenous knowledge is important in farm decision-making to respond to anticipated poor yields. The use of *Senegali* phenology to predict rainfall in the study is also used by Malunga farmers in Tanzania to forecast the upcoming rainy season [25].

The participants' use of celestial bodies to predict rain is corroborated by the use of the moon and the stars by Chibelela farmers. The farmers use the moon's shape and colour as signs to predict a season of either sufficient or scarce rainfall. They also use the movement of stars to make inferences about the rainfall patterns for a specific season of the year [25]. Equally, in Uganda the farmers use local indicators, such as phases and shapes of the moon, to predict upcoming weather [26].

3.2.2. Knowledge of Soil Types

Knowledge of soil varieties by colour and texture, and the types of crops that do well on particular soil types, was evident among the participants. According to the participants, black clayey soil is rich in nutrients and good for cultivation of maize, pumpkin, and gourds. Sandy soil is good for beans, melons, and sweet-reed. Another type of soil is a mixture of sandy and clayey soil which is good for all crops. These findings are supported by observations that Zulu subsistence farmers' knowledge of soil is based on colour and texture of the topsoil, that dark soil indicates higher fertility while lighter soil signifies lower fertility [27].

3.2.3. Mulching

It was reported that previous harvest residue in the form of maize stalks, dried bean and nut plants is a good soil stabiliser. The participants reported that after harvest, the residue is tilled with the soil to improve moisture retention and fertility of the soil. This indigenous practice, according to Buthelezi et al. [27], replenishes depleted soil nutrients.

3.2.4. Soil Fertilization

The participants reported that they apply kraal and poultry manure to make the soil regain fertility, retain moisture, and avoid pests. This type of soil fertilization mainly improves soil moisture conservation [28]. In Tanzania, subsistence farmers understand that if weeds are left to grow, they cover the soil, prevent it from heating up or drying out excessively, induce a positive competition, which simulates crop growth and reduces erosion during rainfall [29].

3.2.5. Seed Selection

The participants reported that, subsequent to harvesting, the crops are threshed and carefully stored for use. The seeds are carefully selected for planting in the next season. Good seeds are selected by colour. Only bright coloured and large-sized seeds are selected for planting. Sometimes selection of the best seeds is done by soaking the seeds in water. Only the sinking seeds are selected and the floating seeds are not selected. Olatokun and Ayanbode [29] observe that Nigerian women cull the seeds and preserve them for the next planting season. In Ethiopia, the farmers select healthy crops in terms of maturity period, height, colour, and size. The panicles or the spikes of the selected varieties are separately harvested, dried, carefully threshed, and the grains are saved for replanting [30].

3.2.6. Multiple Cropping

Sowing of seeds is done haphazardly by hand. All seed varieties are sown simultaneously in the same field. This practice maximises the growth of all crops at the same time in the same field. Inter-planting allows cropping systems to reuse their own stored nutrients [8]. With this system productivity per unit area is higher than in mono-cropping systems with the same level of management. The farmers incorporate a variety of crops with different growth habits in the same field or home-gardens to maximise the chances for production of multiple crops [14].

3.2.7. Maintenance of Crops

It was reported that subsequent to planting, when the crops are about four weeks old, weeding commences. Weeds are removed by hand or hand-hoe to avoid them competing for moisture with the crops, thus disturbing the growth of crops. In Tanzania, when the farmers regard weed competition as negative for crop growth, they perform superficial hoeing, and leave the weeds on the soil surface as protective mulch, to recycle nutrients, and to allow nitrogen assimilation through the bacteria decomposing the plants [31]. For the participants, when the crops are about to reach maturity, the women, boys, and girls spend days in the fields scaring birds off the crops. In many instances a "go-upa" ritual is performed through dispensing of medicine obtained from traditional health practitioners in the field to permanently remove birds and marauding animals from the fields. Olatokun and Ayanbode [29] observe that tobacco (*Nicotiana tabacum*) plants are used to prevent insect build-up on the cocoa plantation. In Indonesia, the farmers burn the common lake-growing plant called *Jariamun* (*Potamogeton. malaianus miq*) in the middle of the rice-field to drive pests from the farm [32].

3.2.8. Storage of Seeds and Crops

After harvesting and threshing, the crops are stored and prevented from attack by weevils. The crops remain fresh until they are all consumed. The most common preservation practice mentioned by the participants is by hanging the maize, sorghum and millet cobs from the hut roof. Sometimes the seeds are mixed with the ash of *Aloe ferox* and stored into clay-pots and baskets. The seeds could last for more than five seasons. Chili pepper (*Capsicum annum.*) is used to preserve harvested cowpea in storage [33].

3.2.9. Fallowing

In many instances exhausted fields are left fallow for two to five years. The participants agreed that this practice helps the soil regain fertility. During fallowing, cattle, sheep and goats are driven in the fields to browse course grass and that their droppings should add to soil fertility. Fallowing enables farmers capture the essence of natural processes of soil regeneration typical of ecological succession [34]. The use of "green manures", which is a recent discovery, intensifies the old fallowing technique in areas where long fallow periods are not possible anymore [34].

4. Conclusions

The study results show that indigenous knowledge is still valuable in the community. The knowledge is embedded in the community's cosmology. Knowledge of plant phenology, and the appearance and shape of the moon and stars is used to plan the planting of crops. The materials used to fertilise the soil, mulch, manage crops, and the seeds are procured at the household level. Soil fertilizers, mulching ingredients and crop management materials are locally developed, always available, affordable, and culture-specific. The study concludes that subsistence farming is sustained by indigenous farming practices and rainfall prediction. The practices involve the improvement of soil structure, maintenance of crops, and the selection and storage of seeds for replanting. Rainfall prediction helps community members plan the planting season. This indigenous knowledge is self-developed and relied upon to generate sustained crop yields to meet subsistence needs. The indigenous knowledge could be helpful towards the achievement of food security at the household level. The knowledge could also make contributions to the development of sustainable adaptation policies to assist rural communities which are vulnerable to climate change hazards.

Acknowledgments: To the members of Dikgale community in Limpopo Province for their kindness and interest in taking part in the study.

Author Contributions: Sejabaledi Agnes Rankoana personally collected and analysed the data presented in this paper. The author has read and approved of the final manuscript.

Conflicts of Interest: The author declares no conflicts of interest.

References

1. Tella, R.D. Towards promotion and dissemination of indigenous knowledge. A case of NIRD. *Int. Inf. Libr. Rev.* **2007**, *39*, 185–193. [CrossRef]
2. Maroyi, A. Enhancing food security through cultivation of traditional food crops in Nhema communal area, Midlands Province, Zimbabwe. *Afr. J. Agric. Res.* **2012**, *7*, 5412–5420.
3. Netting, R.M. *Smallholders, Householders Farm Families and the Ecology of Intensive, Sustainable Agriculture*; Stanford University Press: Stanford, CA, USA, 1993.
4. Zhou, M. Promote conservation and use of underutilized crops. In *Plant Genetic Resources Conservation and Use in China*, Proceedings of the National Workshop on Conservation and Utilization on Plant Genetic Resources, Beijing, China, 5–27 October 2001.
5. Azam-Ali, S. Agricultural diversification: The potential for underutilized crops in Africa's changing climates. *Biol. Forum* **2007**, *100*, 27–38.
6. Sigot, A.J. Indigenous food systems: Creating and promoting sustainable livelihoods. In Proceedings of the International Conference in Indigenous Knowledge Systems: African Perspectives, Thohoyandou, South Africa, 12–15 September 2001.
7. Van Wyk, B.-E. The potential of South African plants in the development of new food and beverage products. *S. Afri. J. Bot.* **2011**, *77*, 857–868. [CrossRef]
8. Miguel, A.; Parviz, K. *Enduring Farms: Climate Change, Smallholders and Traditional Farming Communities*; Third World Network: Penang, Malaysia, 2008.
9. Rook, J. The SADC regional early warning system: Experiences, gains and some lessons learnt from the 1991–92 southern African drought. Usable science: Food security, early warning and El Niño. In Proceedings of the Workshop on ENSO/FEWS, Budapest, Hungary, 25–28 October 1994.
10. Niederwieser, J.D. *Guide to Sweet Potato Production in South Africa*; Roodeplaat vegetable and Ornamental Plant Institute: Pretoria, South Africa, 2004.
11. Nethononda, L.O.; Odhiambo, J.J.O. Indigenous soil knowledge relevant to crop production of smallholder farmers at Rambuda irrigation scheme, Vhembe District South Africa. *Afr. J. Agric. Res.* **2011**, *6*, 2576–2581.
12. Food and Agricultural Organization (FAO). Soaring Food Prices: Facts, Perspectives, Impacts and Actions Required. In Proceedings of the High-Level Conference on World Food Security: The Challenges of Climate Change and Bioenergy, Rome, Italy, 3–5 June 2008.
13. Dube, T.; Phiri, K. Rural livelihoods under stress: The impact of climate change on livelihoods in South Western Zimbabwe. *Am. J. Contemp. Res.* **2013**, *3*, 11–25.
14. Rankoana, S.A. Perceptions of climate change and the potential for adaptation in a rural community in Limpopo Province, South Africa. *Sustainability* **2016**, *8*, 672. [CrossRef]
15. Reid, H.; Huq, S. Mainstreaming community-based adaptation into national and local planning. *Clim. Dev.* **2014**, *4*, 291–292. [CrossRef]

16. Huq, N.; Hugé, J.; Boon, E.; Gain, A.K. Climate change in agricultural communities in rural areas of coastal Bangladesh: A tale of many stories. *Sustainability* **2015**, *7*, 8437–8460. [CrossRef]
17. Polokwane Local Municipality Integrated Development Plan Review, 2011. Available online: http://www.polokwane.gov.za (accessed on 23 June 2016).
18. Statistics South Africa (Statssa) Community Survey: Mid-Year Results 2011. Available online: www.statssa.gov.za (accessed on 15 June 2016).
19. Rankoana, S.A. Aspects of the Ethnobotany of the Dikgale Community in the Northern Province. Master's Thesis in Anthropology, University of the North, Mankweng, South Africa, 2000.
20. Zhu, T.; Ringler, C. *Climate Change Implications for Water Resources in the Limpopo River Basin*; Discussion Paper 00961; International Food Policy Research Institute (IFPRI): Washington, DC, USA, 2010.
21. Food and Agricultural Organization (FAO). *Adaptation to Climate Change in Agriculture, Forestry and Fisheries: Perspective, Framework and Priorities*; Food and Agriculture Organization of United Nations: Rome, Italy, 2009.
22. Speranza, C.I.; Kiteme, B.; Ambenje, P.; Wiesmann, U.; Makali, S. Indigenous knowledge related to climate variability and change: Insights from droughts in semi-arid areas of former Makueni District, Kenya. *Clim. Chang.* **2010**, *100*, 295–315. [CrossRef]
23. Kijazi, A.L.; Chang'a, L.B.; Liwenga, E.T.; Kanemba, A.; Nindi, S.J. The use of indigenous knowledge in weather and climate prediction in Mahenge and Ismani wards, Tanzania. *J. Geogr. Reg. Plan.* **2013**, *6*, 274–280.
24. Chang'a, L.B.; Yanda, P.Z.; Ngana, J. Indigenous knowledge in seasonal rainfall prediction in Tanzania: A case of the south-western highland of Tanzania. *J. Geogr. Reg. Plan.* **2010**, *3*, 66–72.
25. Elia, E.F.; Mutala, S.; Stilwell, C. Indigenous knowledge use in seasonal weather forecasting in Tanzania: The case of semi-arid central Tanzania. *S. Afr. J. Libr. Inf. Sci.* **2014**, *80*, 18–27. [CrossRef]
26. Orlove, B.S.; Roncoli, C.; Kabugo, M.; Majugu, A. Indigenous climate knowledge in southern Uganda: The multiple components of a dynamic regional system. *Clim. Chang.* **2010**, *100*, 243–265. [CrossRef]
27. Buthelezi, N.; Hughes, J.; Modi, A. The Use of Scientific and Indigenous Knowledge in Agricultural Land Evaluation and Soil Fertility of Two Villages in KwaZulu-Natal, South Africa. In Proceedings of the World Congress of Soil Science, Soil Solutions for a Changing World, Brisbane, Australia, 1–6 August 2010.
28. Mapaure, I.; Mhango, D.; Mulenga, K. *Mitigation and Adaptation Strategies to Climate Change*; John Meinert Printing: Windhoek, Namibia, 2011.
29. Olatokun, W.M.; Ayanbode, O.F. Use of indigenous knowledge by rural women in the development of Ogun State. *Indilinga* **2010**, *20*, 47–63. [CrossRef]
30. Belemie, K; Singh, R.K. Conservation of socio-culturally important local crop biodiversity in the Oromia Region of Ethiopia: A case study. *Environ. Manag.* **2012**, *50*, 352–364. [CrossRef] [PubMed]

31. Adedipe, N.O.; Okuneye, P.A.; Ayinde, I.A. The Relevance of Local and Indigenous Knowledge for Nigerian Agriculture. In Presented at the International Conference on Bridging Scales and Epistemologies: Linking Local Knowledge with Global Science in Multi-Scale Assessments, Alexandria, Egypt, 16–19 March 2004.
32. Wahyudy, D.; Anwar, K.; Angelika, A.; Nayu, N.W. Role of Indigenous Knowledge in Traditional Farming System on Natural Resources Management. 2012. Available online: https://www.uni-kassel.de/fb11agrar/fileadmin/datas/fb11/Oekologische_Lebensmittelqualit%C3%A4t_und_Ern%C3%A4hrungskultur/Bilder/David._W_Ploeger_A._Role_of_Indigenous_Knowledge_in_Traditional_Farming_System_on_Natural_Resources_Management.pdf (accessed on 22 September 2016).
33. Altieri, M. Traditional farming in Latin America. *Ecologist* **1991**, *21*, 93–96.
34. Buckles, D.; Triomphe, B.; Sain, G. *Cover Crops in Hillside Agriculture: Farmer Innovation with mucuna*; International Development Research Center: Ottawa, ON, Canada, 1998.

Farmer's Choice of Drought Coping Strategies to Sustain Productivity in the Eastern Cape Province of South Africa

Nomalanga Mary Mdungela, Yonas Tesfamariam Bahta and
Andries Johannes Jordaan

Abstract: This paper determines the factors that influence communal farmers' choice of coping strategies to sustain productivity during drought and to determine current adaptation and coping capacities for drought risk in the Eastern Cape province of South Africa using field surveys, structured and semi-structured interviews and a multinomial probit model. The results identified three main coping strategies used by farmers, namely: irrigation, diversification and drought resistant crops/breeds. On average, most farmers used drought resistant crops/breeds (44%), 32% practiced farm diversification, while 29% used irrigation. Farmers who receive relevant information, have experience and receive sufficient income from their work are more likely to adopt resistant crop varieties and choose suitable animal breeds in case of drought. Access to water has, of course, a significant impact and is positively related to the probability of farmers not adopting any coping strategies. The variable risk level was significant and negatively related to the probability of adopting irrigation as a strategy to address drought. Record keeping was also highly significant and positively associated with the probability of using farm diversification to address drought issues. Education and extension services were not significant. Such viable strategies to reduce the farmer's vulnerability to drought and to improve and sustain productivity should be incorporated into the farmer's existing strategy to adapt and cope with environmental uncertainty. Measures such as rain water harvesting and till practice, keeping reserves, would help them survive through bad years, and increase their agricultural productivity and sustainability.

1. Introduction

Drought is considered as a normal recurring event that affects people around the world and is one of the most important natural disasters in economic, social and environmental terms [1]. Dry periods and drought remain the major meteorological factors that have devastating impacts on the livelihood of the most rural people in South Africa. Ngaka [2] estimated that about 65% of South Africa receives an average annual rainfall of less than 500 mm; this implies that most of the farming in South Africa takes place under arid and semi-arid conditions. In South Africa, drought is a major disaster when considering economic losses and the number of people affected.

73

Akapalu [3] argues that people living in rural areas and resource poor farmers are most vulnerable.

Drought has major implications for the agricultural industry by diminishing production. Every time drought occurs in South Africa, farmers are the most vulnerable as they are the first to be exposed to the devastating effects of drought. According to the Disaster Management Act-57 [4] disaster is declared only when affected people lack the resource capacity to deal with drought.

Vetter [5] found that droughts will pose an increasing challenge to farmers in the future, and finding ways to reduce their ecological and economic impacts should be a major research endeavor. The Red Cross [6] and Mniki [7] focussed on hazard risk and socio-economic factors, which influenced the potential effects of the disaster. Before 1992, the focus was primarily on mitigating the impact of drought on the industrial and commercial agricultural sectors. Jordaan [8] studied drought vulnerability and coping indicators in the Northern Cape Province. Studies on factors that influence communal farmers' choice of coping strategies in the Eastern Cape Province (South Africa) are, however, lacking.

The Eastern Cape Province is the second largest province following the Northern Cape in South Africa and is close to 169,000 square kilometres [9]. The province makes up 13.5% of South Africa's total population. The Eastern Cape Province is made up of 45 municipalities which are grouped as follows: one metropolitan, six district municipalities, and thirty-eight local municipalities. A map of the Eastern Cape Province showing different regions is shown in Figure 1.

The Eastern Cape is one of the six provinces that were declared disaster areas by the previous South African president Thabo Mbeki [10]. The Eastern Cape Province is highly vulnerable to disaster due to a high level of poverty, low standards of living, environmental degradation, poor household economies and a lack of access to resources. The Eastern Cape Province not only has the biggest cattle and sheep herds in South Africa, but is also the place where communal farming is practiced on the largest scale [11].

The following problems were identified among farmers in the Eastern Cape Province of South Africa: insufficient water, animals being injured in accidents, searching for food, receiving early warning information about droughts too late, lack of resources (tractors, land, capital), drought relief does not reach farmers in time and they have to wait for officials from a national department for a disaster declaration. Similar results were found amongst small-holder communal farmers in the Northern Cape [8].

Figure 1. Study areas in the Eastern Cape Province [12].

Hassan and Nhemachena [13] analysed determinants of farm-level climate adaptation measures in Africa using a multinomial choice model fitted to data from a cross-sectional survey of over 8000 farms from 11 African countries. Most studies focus on climate variability and change, adaptive capacity of small scale farmers, farmers' perceptions of drought, cost and risk of coping with drought [14–19]. Less studies deal with factors that influence communal farmers' choice of coping strategies and capacities to drought risk. In view of the importance of the subject and the lack of knowledge with regard to the factors that influence communal farmers' choice of coping strategies, it appears useful to undertake a study.

Thus, the main objective of this study is to determine the factors that influence communal farmers' choices of coping strategies during drought and to determine current adaptation and coping capacities regarding drought risk in the Eastern Cape province of South Africa. To this end, a field survey and multinomial probit model were used. Conclusions drawn could help develop policies and institutional interventions regarding coping strategy and capacities. Moreover, an understanding of factors that influence the choice of communal farmers' coping strategies is critical in designing technological and policy interventions for more effective drought mitigation. This study could not only be applied to South Africa, but to other arid and semi- arid regions as well.

2. Methodology

The study was conducted in three municipalities in the Eastern Cape Province of South Africa, namely Cacadu district municipality, Joe Gqabi district municipality and Oliver Reginald (OR) Tambo district municipality. The focus of this study lay on the communal farmers, where large scale communal farming is practiced and the land is still managed by chiefs or local municipalities. One hundred and twenty-one communal farmers were interviewed from the following districts: Joe Gqabi district municipality (n = 19), Cacadu district municipality (n = 15) and OR Tambo district municipality (n = 87). Primary data was collected by using a structured questionnaire survey and focus group discussion from April–September 2014.

A multinomial probit model was used to investigate the factors that influence the choice of a farmer's specific coping method. According to Munizaga; Daziano and Ziergler [20,21], multinomial probit model applications include constrained and unconstrained versions of the covariance matrix of the multivariate normal distribute of error term. Assumption of a particular covariance structure is unnecessary as the data reveals the substation patterns.

A multinomial probit statistical model is used when there are several possible categories that the dependent variable can fall into. The coping strategies choice model concerns the decision made by farmer "i", i = 1, 2, ... , I of the alternative j in the set $w_i = (1,....,j)$, which produces the highest utility level (V_{ij}). Thus, $V_{i1} < V_{ij}, \forall j \in w_i$ in this notation indicates the choice set is allowed to vary across individuals to account for their own specific coping strategy. The drought coping strategies choices denote to 1 for (none of the strategy), 2 (irrigation), 3 (farm diversification), 4 (resistant crops/breeds) and 5 denote to (more than two coping strategies). Resistant crops/breeds are chosen as base category (option 4). The utilities of other choices (1, 2, 3 and 5) are compared to that of the base category. The individual decision is based on the differences between utility derived from the other drought coping mechanisms and the base category (resistant crops/breeds). This can be represented as:

$$Y^*{}_{ij} = V_{ij} - V_{ij} \qquad (1)$$

where $Y^*{}_{ij}$ denotes an unobservable choice made, when individual i chooses option j. If $Y^*{}_{ij} < 0$ for $j = 1, ..., J$, then Y_i farmer I chooses the base category option (drought resistant crops/breeds) and $Y_i = 0$. Otherwise, farmer i's choice yields the a higher value for $Y^*{}_{ij}$ and $Y_{ij} = j$. Assuming that each farmer i faces the same j alternatives, a multinomial probit model formulation based on linear-in-parameters utilities may be written as follows:

$$V_{ij} = Z_{ij}\beta + \varepsilon_{ij}, \varepsilon_{ij} \quad N(0, \textstyle\sum) \qquad (2)$$

$$y_{ij} = \begin{cases} 1|if\ V_{ij} \le V_{ij}\ for\ i\ =\ 1,\ ...,\ I;\ j\ =\ 1,\ ...,\ J \\ 0|otherwise \end{cases} \qquad (3)$$

The variable Y_{ij} denotes the choice made by farmer i, V_{ij} is the unobservable utility of alternative j as perceived by individual i, Z_{ij} is a $(1 \times K)$ vector explanatory variables characterizing both alternative j and the individual i. β is a $(K \times 1)$ vector of fixed parameters, and, finally, ε_{ij} is a normally distributed random error term of mean zero assumed to be correlated with the errors associated with the other alternatives $j, j = 1,..., J, j \neq i$; and covariance matrix of:

$$\Sigma = Cov(\varepsilon_i = \begin{pmatrix} \sigma_{11} & \sigma_{12} & \sigma_{13} \\ \sigma_{21} & \sigma_{22} & \sigma_{23} \\ \sigma_{31} & \sigma_{32} & \sigma_{33} \end{pmatrix}) \tag{4}$$

with $\sigma_{ij} > 0, \forall_j$(positive definiteness). The predicated probability of choosing any of the coping strategies choices represented with the following Equations (5)–(9):

$$P(y_i = 1)P(V_{i1} + \varepsilon_{i1} > V_{i2} + \varepsilon_{i2} \text{ and } V_{i1} + \varepsilon_{i1} > V_{i3} + \varepsilon_{i3}) \tag{5}$$

$$P(y_i = 2)P(V_{i2} + \varepsilon_{i2} > V_{i1} + \varepsilon_{i1} \text{ and } V_{i2} + \varepsilon_{i2} > V_{i3} + \varepsilon_{i3}) \tag{6}$$

$$P(y_i = 3)P(V_{i3} + \varepsilon_{i3} > V_{i1} + \varepsilon_{i1} \text{ and } V_{i3} + \varepsilon_{i3} > V_{i2} + \varepsilon_{i2}) \tag{7}$$

$$P(y_i = 4)P(V_{i4} + \varepsilon_{i4} > V_{i1} + \varepsilon_{i1} \text{ and } V_{i4} + \varepsilon_{i4} > V_{i5} + \varepsilon_{i5}) \tag{8}$$

$$P(y_i = 5)P(V_{i5} + \varepsilon_{i5} > V_{i1} + \varepsilon_{i1} \text{ and } V_{i5} + \varepsilon_{i5} > V_{i4} + \varepsilon_{i4}) \tag{9}$$

Assuming that the response categories are mutually exclusive and exhaustive, then $\sum_{j=1}^{J} P_{ij} = 1$. For each i, the probabilities add up to one for each individual and we have only $J-1$ parameters. This implies that Equation (5) + (6) + (7) + (8) + (9) = 1, which is rewritten as:

$$P(y_i = 1) + P(y_i = 2) + P(y_i = 3) + P(y_i = 4) + P(y_i = 5) = 1 \tag{10}$$

Multinomial probit was adopted to avoid the limitations of the simpler multinomial logit, i.e., nonsensical predictions, since the dependent variable is not continuous, recoding the dependent variable can give different results [22]. Multinomial probit estimates correlation, depending upon choice size and if correlation is high. Therefore, multinomial probit is designed to be used only if the options are relatively small [23]. Empirically, the multinomial probit regression can be written as follows:

$$L_{ij} = \alpha_{0ij} + \alpha_{ij}D_{ij} + \alpha_{2ij}K_{ij} + \alpha_{3ij}Fs_{ij} + \\ \alpha_{4ij}Aw_{ij} + \alpha_{5ij}Al_{ij} + \alpha_{6ij}Es_{ij} + \alpha_{7ij}Rk_{ij} + \alpha_{8ij}Ed_{ij} + \alpha_{9ij}I_{ij} + e_{ij} \tag{11}$$

where ij denotes coping strategies ($j = 1$ denotes no coping strategies, $j = 2$ denotes irrigation, $j = 3$ denotes farm diversification, $j = 4$ denotes drought resistant crops/breeds and $j = 5$ more than one coping strategy). $Dij = 1$ if farmer i received information from the Department of Agriculture and Rural Development; K denotes knowledge of a farmer and Fs represents farming experience. $Aw = 1$ if farmer has access to water and $Al = 1$ denotes access to land. $Es = 1$ if farmer received extension services. *Ed* denotes educational level (primary, high school or degree). *Rk* = 1 if farmer keeps records. *I* denotes income from farming activities. α_{oij} denotes the constant term and $\alpha_{1ij}, \alpha_{2ij}, \ldots, \alpha_{10ij}$ represent the coefficients of the explanatory variables in the model, while e_{ij} denotes the disturbance term.

3. Results and Discussion

3.1. Demographic and Socio Economic Characteristics of the Respondents

Some of the respondents' socioeconomic characteristics are provided in Table 1. More males (73%) than females (27%) took part in the study. A possible reason for the male dominated farming activities in the study area might be that they have access to land. Quisumbing [24] reported that there is a great disparity in the size of landholdings between men and women, and that the mode of women participation in agricultural production varies with the land-owning status of households.

Table 1. Socio-economic characteristics of the respondents.

Characteristics	Sub-Characteristics	OR Tambo (n = 87)		Joe Gqabi (n = 19)		Cacadu (n = 15)		% (N = 121)			Total
		n	%	n	%	n	%	ORT	JG	CD	%
Age (yrs)	25–34	7	8	3	16	2	13	6	2	2	10
	35–44	20	23	3	16	2	13	17	2	2	21
	45–54	25	29	4	21	4	27	21	3	3	27
	>55	35	40	9	47	7	47	29	7	6	42
Gender	Male	62	71	16	84	11	73	51	13	9	73
	Female	25	29	3	16	4	27	21	2	3	27
Education	None	23	26	1	5	3	28	19	1	2	22
	Primary	44	51	13	68	12	80	36	11	10	57
	Secondary	18	21	2	11	—	—	15	2	—	17
	Graduate	2	2	3	16	—	—	2	2	—	4
Household size	0–4	29	33	5	26	8	54	24	4	7	35
	5–8	32	37	11	58	5	33	26	9	4	39
	9–12	14	16	3	16	2	13	12	2	2	16
	>13	12	14	—	—	—	—	10	—	—	10
Access to resources	Land	69	79	17	89	14	93	57	14	12	83
	Water	35	40	10	53	6	40	29	8	5	42
Experience (yrs)	0–4	10	12	4	21	7	47	8	3	6	17
	5–9	20	23	6	32	4	27	17	5	3	25
	10–14	28	32	4	21	2	13	23	3	2	28
	>15	29	33	5	26	2	13	24	4	2	30

ORT = OR Tambo; JG = Joe Gqabi; CD = Cacadu district municipality.

Many respondents (22%) did not have a formal education, 17% had a secondary level education and few (4%) had a tertiary education. The demographic and socio-economic characteristics are important, because they influence households' economic behaviour [25].

Most (83%) of the communal farmers had access to resources (land) and of these, 57% were in OR Tambo district municipality, 14% in Joe Gqabi district municipality and the rest 12% in Cacadu district municipality. Forty-two (42%) communal farmers had access to water area, the majority (29%) were in OR Tambo district municipality. Forty-two (42%) of the respondent were 55 years or older, 39% had household sizes of between 5 and 8 people, and 58% of the respondents had more than 10 years' experience in farming.

3.2. Determinants of the Choice of Drought Coping Strategies

The multinomial probit regression model was used to examine the factors that influence the choice of communal farmers' coping strategies during drought in the Eastern Cape Province. Table 2 represents maximum likelihood estimates of the multinomial probit regression model (The detailed results of the multinomial probit regression model are available in Appendix A, Table A1). Drought resistant crops/breeds were used as reference category for the multinomial probit analysis, because most farmers opted for it. Income, experience, access to land and water, and information from the Department of Agriculture and Rural Development (DARD) were variables fitted in the model, because of significant influence the choice of coping mechanisms.

The coefficient for DARD information is negatively related to the probability of farmers not adopting any coping strategies and was highly significant at the 1% level. This implies that farmers who received information from the DARD are more likely to adopt resistant crop varieties and animal breeds rather than adopting no drought coping strategies. Information on earlier drought impact is very important for planning future drought responses. By comparing most severe impacts of drought, policymakers can plan to minimise the most severe impacts [26,27].

The coefficient for access to water is significant at the 5% level and is positively related to the probability of farmers not adopting any coping strategies. This result is plausible, because the farmers who have access to water already have mitigation strategies to address drought, therefore there might be no need for them to adopt any other strategies. Communal farmers have access to water, but there was insufficient and this limits their ability to expand their farming businesses.

The coefficient for access to land had a positive association with the likelihood of choosing irrigation in favour of resistant crops/breeds with a significance of 1%. This indicates that farmers are more likely to engage in irrigation especially in the Cacadu municipality, because 47% of respondents indicated irrigation use. Most of these farmers produce crops and vegetables, which requires more water compared to other districts. Contrarily, access to land was significant at the 10% level and negatively related to not adopting any of the drought coping strategies. This suggests that farmers are more likely to adopt drought resistant crop varieties or animal breeds, which will curtail the effects of climate change on their production. Previous studies found that farmers having secure land tenure were likely to take up adaptation strategies [28,29].

Table 2. Multinomial probit regression analysis.

Variables	No Coping Strategy		Irrigation		Farm Diversification		More Than Two Coping Strategy	
	Coef.	$p > |z|$	Coef.	$p > |z|$	Coef.	$p > |z|$	Coef.	$p > |z|$
DARD	−1.954	0.002 ***	−0.753	0.247	0.357	0.59	0.299	0.597
Knowledge	0.942	0.159	−0.449	0.529	0.367	0.571	0.657	0.235
Agricultural training	−0.214	0.734	−1.219	0.121	−0.705	0.296	−0.346	0.559
Experience	−0.042	0.046 **	−0.094	0.043 **	−0.011	0.654	−0.003	0.952
Access to land	1.309	0.065 *	3.602	0.000 ***	0.7	0.331	0.6	0.227
Access to water	1.421	0.024 **	−0.333	0.586	0.674	0.203	0.389	0.424
Risk level	−0.499	0.296	−0.019	0.014 *	0.902	0.365	−0.085	0.667
Extension services	0.907	0.165	−0.734	0.364	−0.859	0.222	−0.654	0.286
Farmers associations	−1.362	0.0018 **	−42.73	—	−1.882	0.007 ***	−1.084	0.044 **
Record keeping	0.392	0.497	2.533	0.000 ***	1.889	0.001 ***	0.484	0.351
Education2	−0.983	0.274	0.285	0.758	0.959	0.293	−1.086	0.201
Education 3	0.41	0.524	0.781	0.392	−0.318	0.613	0.799	0.172
Education 4	42.78	—	1.109	0.345	0.242	0.738	−0.438	0.53
Income 2	−1.288	0.149	1.272	0.081 *	2.601	0.000 ***	2.03	0.006 *
Income 3	−1.621	0.093*	3.823	—	4.082	—	5.3	0.000 ***
Base category	Drought resistant crop or animal breed							
Number of observations	—	—	—	—	—	—	—	121
Wald chi2(76)	—	—	—	—	—	—	—	0.000
Log pseudolikelihood	—	—	—	—	—	—	—	−108.56
Prob> chi2	—	—	—	—	—	—	—	0.000

***, **, * = significant at 1%, 5%, and 10% probability level, respectively.

The coefficient of experience correlated with the probability of not adopting drought coping strategies in favour of adopting drought resistant crops or animal breeds with a 5% significance. The negative sign of experience implies that farmers who have been in agricultural production are more likely to adopt drought resistant crops or animal breeds to mitigate climate change. Experienced farmers have gathered enough information on weather patterns over a period of time and will therefore be able to choose the appropriate means to address changing weather patterns. Similarly, experience negatively correlated with the probability of adopting irrigation as a mitigating drought strategy in favour of drought resistant crops or animal breeds. The result implies that experienced farmers are more likely to adopt irrigation as a drought coping strategy. Developing irrigation facilities may be costly compared to using drought resistant crops or animal breeds. This implies that communal farmers are more vulnerable as they do not receive enough income due to drought impacts. Studies show that the greater the experiences in agricultural farming, the more likely farmers are to have good knowledge about the weather and climatic conditions and thus adapt. Hisali et al. [28] pointed to the importance of farming experience in adaptation decision making.

The coefficient of income 3 (South African Rand ZAR 50,000–ZAR 100,000) was significant at the 10% level and had negative effects on the likelihood of not adopting any drought coping strategies in favour of drought resistant crops or animal breeds. The results suggests that farmers with income ranges between ZAR 10,000 and ZAR 50,000 per annum are more likely to adopt drought resistant crop or animal breeds as mitigating strategy against drought. The reason is that these farmers have less income and are more vulnerable to drought and thus more food insecure. Farmers with income 4 (ZAR 100,000–ZAR 200,000) can afford to purchase drought resistant crops or animal breeds. Moreover, farmers with income level 5 (> ZAR 200,000) are more likely to adopt more than one drought coping strategy than only drought resistant crops or animal breeds. For example, money can be used to buy additional feed for livestock to survive until the dry period is over.

The variable risk level was significant at 10% and negatively related to the probability of adopting irrigation as a strategy to address drought. This implies that farmers with higher risk levels are less likely to adopt irrigation as a coping strategy. However, risk levels were positively related to the probability of using farm diversification as a drought mitigating strategy in favour of drought resistant crops or animal breeds. Farm diversification may help farmers to cope better during drought as they have additional crops or livestock to support their main farming enterprises. The level of perceived risk associated with the capacity to adapt to climate change determines the likelihood of adopting adaptation measures [28].

Farmers associations are significant at 1%, meaning a probability that those farmers who receive information from associations have the higher probability of adopting farm diversification. However, this is not happening as associations do not operate at grass root level.

Record keeping was highly significant at 1% and positively associated with the probability of using farm diversification to address drought issues. Record keeping helps the farmer track climate patterns and the performance of the farming operation and enables the farmer to explore alternative risk reduction methods associated with drought.

Even though education and extension services were not significant, they negatively influenced the farmers' probability to use one of the coping strategies. This suggests that farmers can use their education with extension services/support to make informed decisions about their farming. Furthermore, it reduces their level of vulnerability to drought when using coping mechanisms. Education is one of the key determinants in adopting adaptation strategies. Higher education level increases the individual awareness of different alternatives [30].

3.3. Determine Current Adaptation, Coping Capacities and Strategies Opted by Communal Farmers

Coping strategies are remedial actions undertaken by those whose livelihoods are threatened. This involves managing resources both during drought and in normal times in order to withstand the effects of drought risk. Irrigation, farm diversification and drought resisting coping capacities and strategies were selected based on literature, expert opinion, observations, level of ease of measurement and their importance. It was found that 44% of the farmers used drought resistant crops or animal breeds, 32% practiced farm diversification and 29% used irrigation on their farms. Farmers found it easier to farm with drought resistant crops or animal breeds due to the nature of drought tolerance than with irrigation coping strategies. Eriksen et al. [31] describe coping mechanisms as the actions and activities that take place within existing structures and systems; e.g., the introduction of on-farm diversification. Diversification can include alternative feed and fodder sources or livestock types [32].

Figure 2 displays that more OR Tambo farmers employ cultivars or breeds that are less sensitive to drought compared to other two districts. Cacadu district farmers use more irrigation; therefore, they are less vulnerable during drought as they are likely to diversify because of high level of water availability. Figure 2 also indicates that only few farmers employ diversification in their farming activities, which leaves most of the farmers vulnerable. Nevertheless, farmers who practice diversification on their farms are more resilient during drought. Farmers can farm with ostrich and goat since these are recognised to have a potential in the Eastern Cape Province.

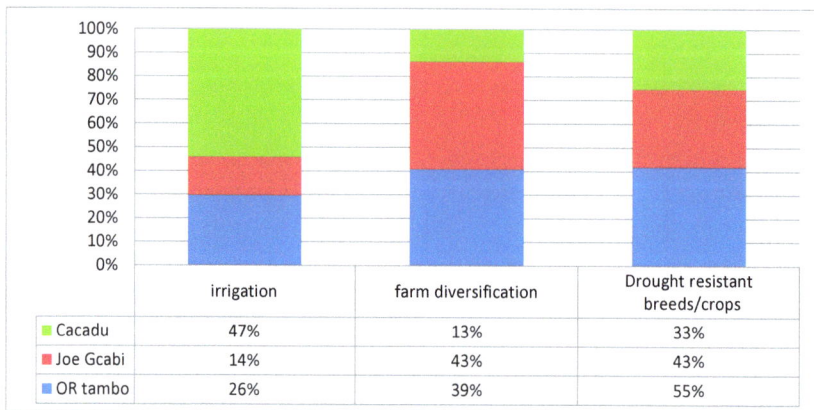

	irrigation	farm diversification	Drought resistant breeds/crops
Cacadu	47%	13%	33%
Joe Gcabi	14%	43%	43%
OR tambo	26%	39%	55%

Figure 2. Coping mechanisms used by farmers in the Study area.

To manage drought effectively, diversifying livelihood strategies and income generating options within and outside agriculture is required, especially through non-farm enterprises and employment opportunities. Off-farm income in the three districts should be explored during drought. Farmers also plant food gardens to support their families, others keep chickens.

Communal farmers sell their excess animals and non-farming assets to buy feed for their livestock. Other farmers plant oats to make silage and lucerne for grazing which can be used in dry periods. The farmers were willing to change the type of livestock, crops and reduce herd sizes. It was argued by extension officers that lucerne can be used for fodder banks, but farmers found it too expensive to produce.

The study found that these coping mechanisms assist farmers to cope during drought. Drought insurance and/or tax free savings schemes can be used as tools to increase coping capacity and requires further research.

4. Conclusions

This study investigated the factors that influence the choice of communal farmers' coping strategies in OR Tambo, Joe Gqabi and Cacadu districts of the Eastern Cape Province of South Africa. Firstly, the results indicate that the choice of communal farmers' coping strategies should take into consideration farmers' access to land, income, experience, and education. Non-consideration of factors can lead to the choice of inappropriate coping strategies.

Secondly, the vulnerability and limited coping capacities of farmers is highly correlated with the inability to access land, water, finance, market and timely information. In South Africa, drought has significant negative impacts and continues to pose long lasting effects on the agricultural sector. It was noted that the majority of farmers had limited abilities to deal with drought issues due to a lack of access to resources and information.

Thirdly, any viable strategy to reduce the farmers' vulnerability to drought and to improve productivity should be incorporated into existing adaptation strategies regarding environment uncertainty. Measures such as rain water harvesting, tilling practices, and keeping reserves help farmers to survive and increase their agricultural productivity. Moreover, disaster risk management committees from different stakeholders at municipality level could be established and, together with extension services, improve early warning and information systems. This must be effectively communicated to the communities before disaster strikes. A holistic approach between different stakeholders should be practiced.

Generally, the findings highlight the need for policy adjustments with regard to drought impact, vulnerabilities and lack of coping capacities that take into consideration communal farmers' existing coping strategies as well as the factors that influence the choice of coping strategy.

Acknowledgments: We would like to thank the Water Research Commission (WRC) and Department of Agriculture, Forestry and Fisheries (DAFF) for their support and funding for this project.

Author Contributions: Nomalanga Mary Mdungela was responsible for collecting data for the multinomial probit regression model, the writing of this article and assisted with the application of the multinomial probit regression model. Yonas Tesfamariam Bahta and Andries Johannes Jordaan assisted in technical aspects of writing, data collection, editing and the application of the multinomial probit regression model. All authors have read and approved the final manuscript.

Conflicts of Interest: The authors declare no conflict of interest.

Appendix A.

Table A1. Multinomial probit regression analysis.

Multinomial Probit Regression Log Pseudolikehood = −108.5604				$N =$ Wald chi2(76) Probability > chi2		121 0.0000 0.0000	
Variables	Coefficient	Robust Standard Error	z	$p > |z|$	(95% CI)		
1 (no strategy)							
DARD	−1.9542	0.6353	−3.08	0.002	−3.1994	−0.7090	
Private sector	0.9709	0.5756	1.69	0.092	−0.1574	2.0991	
Knowledge	0.9423	0.6695	1.41	0.159	−0.3699	2.2545	
Agricultural training	−0.2148	0.6329	−0.34	0.734	−1.4552	1.0257	
Indigenous knowledge	−0.6139	0.5859	−1.05	0.295	−1.7622	0.5345	
Experience	−0.0429	0.0216	−1.95	0.046	−0.0852	−0.0007	
Access to land	1.3092	0.7097	1.84	0.065	−0.0817	2.7001	
Access to water	−1.4209	0.6304	2.25	0.024	0.1853	2.6566	
Level of debt	−1.0506	0.7255	−1.45	0.148	−2.4728	0.3714	
Risk level	0.2647	0.1937	1.37	0.172	−0.1149	0.6443	
Extension services	−0.9069	0.6526	−1.39	0.165	−2.1860	0.3723	
Farm associations	−1.3618	0.5743	−2.37	0.018	−2.4875	−0.2361	
Record keeping	0.3915	0.5758	0.68	0.497	−0.7370	1.5200	
Education 2	−0.9826	0.8988	−1.09	0.274	−2.7442	0.7790	
Education 3	0.4103	0.6439	0.64	0.524	−0.8517	1.6723	
Education 4	0.6261	0.7762	0.81	0.420	−0.8953	2.1474	
Education 5	−42.7807	—	—	—	—	—	
Income 2	−1.2883	0.8924	−1.44	0.149	−3.0373	0.4608	
Income 3	−1.6212	0.9663	−1.68	0.093	−3.5152	0.2727	
Income 4	−32.33932	—	—	—	—	—	
2 (irrigation)							
DARD	−0.7533	0.6505	−1.16	0.247	−2.0283	0.5218	
Private sector	−0.0098	0.6611	−0.01	0.988	−1.3055	1.2859	
Knowledge	−0.4498	0.7147	−0.63	0.529	−1.8507	0.9511	
Agricultural training	−1.2185	0.7865	−1.53	0.121	−2.7599	0.3230	
Indigenous knowledge	−0.6964	0.5818	−1.20	0.231	−1.8367	0.4439	
Experience	−0.0938	0.0463	−2.03	0.043	−0.1845	−0.0031	
Access to land	3.6015	0.9826	3.67	0.000	1.6756	5.5274	
Access to water	−0.3335	0.6117	−0.55	0.586	−1.5325	0.8654	
Level of debt	0.1356	0.6973	0.19	0.846	−1.2311	1.5023	
Risk level	−0.4986	0.2961	−1.68	0.092	−1.0789	0.0817	
Extension services	0.7340	0.8083	0.91	0.364	−0.8502	2.3182	
Farm associations	−42.7349	—	—	—	—	—	
Record keeping	2.5334	0.6671	3.80	0.000	1.2258	3.8409	
Education 2	0.2847	0.9238	0.31	0.758	−1.5260	2.0954	
Education 3	0.7810	0.9132	0.86	0.0392	−1.0089	2.5710	

Multinomial Probit Regression Log Pseudolikehood = −108.5604				*N* = Wald chi2(76) Probability > chi2		121 0.0000 0.0000
Variables	Coefficient	Robust Standard Error	*z*	*p* > |*z*|	(95% CI)	
Education 4	1.1093	1.1738	0.95	0.345	−1.1913	3.4098
Education 5	−0.6677	0.9370	−0.71	0.476	−2.5042	1.1688
Income 2	1.2716	0.7297	1.74	0.081	−0.1586	2.7019
Income 3	3.8234	—	—	—	—	—
Income 4	−22.9572	—	—	—	—	—
3 (farm diversification)						
DARD	−0.3569	0.6615	−0.54	0.590	−1.6535	0.9397
Private sector	1.1449	0.6413	1.79	0.074	−0.1121	2.4018
Knowledge	0.3674	0.6449	0.57	0.571	−0.9049	1.6396
Agricultural training	−0.7046	0.6737	−1.05	0.296	−2.0249	0.6158
Indigenous knowledge	−1.1342	0.5459	−2.08	0.038	−2.2041	−0.0644
Experience	−0.0108	0.0241	−0.45	0.654	−0.0579	.0364
Access to land	0.7003	0.7200	0.97	0.331	−0.7108	2.1115
Access to water	0.6735	0.2167	1.27	0.203	−0.3636	1.7105
Level of debt	1.3780	0.7037	2.06	0.039	0.0690	2.6870
Risk level	−0.0193	0.2167	−0.09	0.929	−0.4441	0.4055
Extension services	−0.8596	0.7037	−1.22	0.222	−2.2388	0.5196
Farm associations	−1.8816	0.6920	−2.72	0.007	−3.2378	−0.5253
Record keeping	1.8894	0.5761	3.28	0.001	0.7601	3.0186
Education 2	−0.9587	0.9110	−1.05	0.293	−2.7443	0.8269
Education 3	−0.3175	0.6278	−0.51	0.613	−1.5480	0.9130
Education 4	0.2418	0.7826	0.33	0.738	−1.1775	1.6611
Education 5	0.3997	0.8626	0.46	0.643	−1.2911	2.0904
Income 2	2.6012	0.5964	4.36	0.000	1.4324	3.770
Income 3	4.0823	—	—	—	—	—
Income 4	21.66639	—	—	—	—	—
5 (more than two coping strategies)						
DARD	0.2989	0.5653	0.53	0.597	−0.8091	1.4070
Private sector	0.4729	0.5290	0.09	0.929	−0.9894	1.0840
Knowledge	0.6571	0.5528	1.19	0.235	−0.4263	1.7405
Agricultural training	−0.3456	0.5915	−0.58	0.559	−1.5050	0.8138
Indigenous knowledge	−0.5770	0.4865	−1.19	0.236	−1.5305	0.3765
Experience	−0.0026	0.0278	−0.09	0.925	−0.0571	0.0518
Access to land	0.6000	0.5515	1.09	0.277	−0.4809	1.6809
Access to water	0.3899	0.4875	0.80	0.424	−0.5656	1.3454
Level of debt	0.3453	0.5640	0.61	0.540	−0.7601	1.4507
Risk level	0.1654	0.1781	0.93	0.353	−0.1837	0.5144
Extension services	−0.6538	0.6123	−1.07	0.286	−1.8539	0.5463
Farm associations	−1.0839	0.5389	−2.01	0.044	−2.1401	−0.0276
Record keeping	0.4837	0.5185	0.93	0.351	−0.5326	1.5001
Education 2	−1.0856	0.8497	−1.28	0.201	−2.7510	0.5797
Education 3	0.7997	0.5857	1.37	0.172	−0.3482	1.9477
Education 4	−0.4378	0.6976	−0.63	0.530	−1.8051	0.9296
Education 5	−0.7915	0.8666	−0.91	0.361	−2.4900	0.9071
Income 2	−2.0301	0.7457	2.72	0.006	0.5685	3.4917
Income 3	5.300	0.9555	5.55	0.000	3.4270	7.1724
Income 4	21.7229	—	—	—	—	—

CI = confidence intervals; DARD = Department of Agriculture and Rural Development

References

1. Buckland, R.; Eele, G.; Mugwara, R. Humanitarian crisis and natural disasters: A SADC perspective. In *Food Aid and Human Security*; Routledge Research EADI (European Association of Development Research and Training Institute) Studies in Development 24; Clay, E., Stokke, O., Eds.; Taylor and Francis: Abingdon-on-Thames, UK, 2000; pp. 150–181.
2. Ngaka, M.J. Drought preparedness, impact and response: A case of the Eastern Cape and Free State provinces of South Africa. *J. Disaster Risk Studies* **2012**, *4*, 1–6. [CrossRef]
3. Akapalu, D.A. Response Scenarios of Households to Drought-Driven Food Shortage in a Semi-Arid Area in South Africa. Master's Thesis, University of the Witwatersrand, Johannesburg, Gauteng, South Africa, 2005.
4. Republic Of South Africa. Disaster Management Act-57, 2002. Government of South Africa. Available online: https://www.westerncape.gov.za/Text/2004/10/a57-02.pdf (accessed on 22 December 2016).
5. Vetter, S. Drought, change and resilience in South Africa's arid and semi-arid rangelands. *S. Afr. J. Sci.* **2009**, *105*, 29–33. [CrossRef]
6. Red Cross. *World Disasters Report: Focus on Reducing Risk*; International Federation of Red Cross and Red Crescent Societies: Bloemfield, CT, USA, 2002.
7. Mniki, S. Socio-Economic Impact of Drought Induced Disasters on Farm Owners of Nkonkobe Local Municipality. Master's Thesis, University of the Free State, Bloemfontein, South Africa, 2009.
8. Jordaan, A.J. Drought Risk Reduction in the Northern Cape Province, South Africa. Ph.D. Thesis, University of the Free State, Bloemfontein, South Africa, 2012.
9. Hamann, M.; Tuinder, V. *Introducing the Eastern Cape: A Quick Guide to Its History, Diversity and Future Challenges*; Stockholm Resilience Centre, Research for Governance of Social-Ecological Systems, Stockholm University: Stockholm, Sweden, 2012.
10. IRIN (Integrated Regional Information Networks) South Africa: Drought emergency in six provinces affect 4 million. *IRIN News*. 19 January 2014. Available online: http://www.irinnews.org/news/2004/01/19/drought-emergency-six-provinces-affects-4-million (accessed on 22 December 2016).
11. Nowers, C. Stock Farming: What Is Holding Back Communal Sheep and Cattle Production? *Farming S. Afr.* **2008**, 26–27.
12. Council for Scientific and Industrial Research (CSIR) Geospatial Analysis Platform (GAP). Available online: http://www.gap.csir.co.za/images/images/GAPmesozones2010.pdf/view (accessed on 25 January 2015).
13. Hassan, R.; Nhemachena, C. Determinants of African farmers' strategies for adapting to climate change: Multinomial choice analysis. *Afr. J. Agric. Resour. Econ.* **2008**, *2*, 83–104.
14. Cooper, P.J.M.; Dimes, J.; Rao, K.P.C.; Shapiro, B.; Shiferaw, B.; Twomlow, S. Coping better with current climatic variability in the rain-fed farming systems of sub-Saharan Africa: An essential first step in adapting to future climate change? *Agric. Ecosyst. Environ.* **2008**, *126*, 24–35. [CrossRef]

15. Mertz, O.; Mbow, C.; Reenberg, A.; Diof, A. Farmers' Perceptions of Climate Change and Agricultural Adaptation Strategies in Rural Sahel. *Environ. Manag.* **2009**, *43*, 804–816. [CrossRef] [PubMed]

16. Meza, L.E.R. Adaptive capacity of small-scale coffee farmers to climate change impacts in the Soconusco region of Chiapas, Mexico. *Clim. Dev.* **2015**, *7*, 100–109. [CrossRef]

17. Roncoli, C.; Ingram, K.; Kirshen, P. The costs and risks of coping with drought: livelihood impacts and farmers' responses in Burkina Faso. *Clim. Res.* **2001**, *19*, 119–132. [CrossRef]

18. Takasaki, Y.; Rarham, B.L.; Coomes, O.T. Risk coping strategies in tropic forest, floods, illness and resources extraction. *Environ. Dev. Econ.* **2004**, *2*, 203–224. [CrossRef]

19. Twomlow, S.; Mugab, F.T.; Mwale, M.; Delve, R.; Nanja, D.; Carberry, P.; Howden, M. Building adaptive capacity to cope with increasing vulnerability due to climatic change in Africa—A new approach. *Phys. Chem. Earth* **2008**, *33*, 780–787. [CrossRef]

20. Munizaga, M.A.; Daziano, A. Testing mixed logit and probit by simulation. Transportation research Record'. *J. Transp. Res. Board* **2005**, *1921*, 53–62. [CrossRef]

21. Ziergler, A. Individual characteristics and stated preferences for alternative energy sources and propulsion technologies in vehicles: A discrete choice analysis. *Transportation Research Part A: Policy Pract.* **2012**, *46*, 1372–1385.

22. Rosella, L.; Walton, R. Multinomial logistic regression: Analysis of multi-category outcomes and its application to a Salmonella Enteritidis investigation in Ontario. 2013. Available online: www.publichealthontario.ca/en/LearningAndDevelopment/Events/Documents/Multinomial_logistic_regression_2013.pdf (accessed on 12 January 2015).

23. Donkor, E.; Owusu, V. Examining the socioeconomics determinants of rice farmer's choice of land tenure systems in the upper east region of Ghana'. *J. Agric. Technol.* **2014**, *10*, 505–515.

24. Quisumbing, A. *Gender Differences in Agricultural Productivity: A Survey of Empirical Evidence*; Discussion Paper Series No. 36; Education and Social Policy Department, World Bank: Washington, DC, USA, 1994.

25. Randela, R. Integration of Emerging Cotton Farmers into the Commercial Agricultural Economy. Master's Thesis, University of the Free State, Bloemfontein, South Africa, 2005.

26. Dziegielewski, B.; Garbharran, H.P.; Langowski, J.F. Lessons learned from the California drought (1987–1992). In *National Study of Water Management during Drought: A Research Assessment*; IWR Report 91-NDS-3; Institute for Water Resources, US Army Corps of Engineers: Alexandria, VA, USA, 1991.

27. Udmale, P.; Ichikawa, Y.; Manandhar, S.; Ishidaira, H.; Kiem, A.S. Farmers' perception of drought impacts, local adaptation and administrative mitigation measures in Maharashtra State, India. *Int. J. Disaster Risk Reduct.* **2014**, *10*, 250–269. [CrossRef]

28. Hisali, E.; Birungi, P.; Buyinza, F. Adaptation to climate change in Uganda: Evidence from micro level data. *Glob. Environ. Chang.* **2011**, *21*, 1245–1261. [CrossRef]

29. Deressa, T.T.; Hassan, R.M.; Ringler, C.; Alemu, T.; Yesuf, M. Determinants of farmers' choice of adaptation methods to climate change in the Nile Basin of Ethiopia'. *Glob. Environ. Chang.* **2009**, *19*, 248–255. [CrossRef]

30. Alam, K. Farmers' adaptation to water scarcity in drought-prone environments: A case study of Rajshahi District, Bangladesh'. *Agric. Water Manag.* **2015**, *148*, 196–206. [CrossRef]

31. Eriksen, S.; Brown, K.; Kelly, P.M. The Dynamics of Vulnerability: Locating coping strategies in Kenya and Tanzania. *Geogr. J.* **2005**, *171*, 287–305. [CrossRef]

32. Jordaan, A.J.; Sakulski, D.; Jordaan, A.D. Interdisciplinary drought risk assessment for agriculture: The case of communal farmers in the Northern Cape Province, South Africa. *South Afr. J. Agric. Ext.* **2013**, *41*, 1–16.

Applying a Decision Support Model to Investigate the Influence of Precision Agriculture Practices on Sustainable Crop Production

Frikkie Alberts Maré and Hendrik Petrus Maré

Abstract: The concepts of precision agriculture (PA) and sustainability are inextricably linked. PA can be described as a catch-all term for techniques, technologies, and management strategies that address in-field variability. Sustainable agriculture, in short, strives to enhance environmental quality and the resource base on which agriculture depends. The main objective of the study is to investigate the impact of PA practices on the sustainability of a crop production enterprise in comparison with conventional farming (CF). The procedures that were used to achieve the objective included the scanning of fields with a gamma-ray spectrometer for the identification of management zones and the application of a decision support model, namely the Scenario Planning, Analysis and Risk Evaluation (SPARÉ) model, to investigate the impact of precision agriculture practices on sustainability. Three crops—maize, wheat, and soya beans—were used in the model to generate the results. The results of the study indicate that precision agriculture does enhance sustainability, as the amount of lime and gypsum, fertiliser, and water that are applied per ton of grain harvested decrease by 22.6%, 11.9%, and 24.1%, respectively, on average for the three crops, making the resource use more sustainable than with conventional agriculture. The gross margin of the whole farm scenario increased with 26.9% and, thus, increased the financial sustainability of the whole farm enterprise.

1. Introduction

Sustainable agriculture has been defined in numerous different ways in the past, and it is especially due to the debated meaning of "sustainability" that different definitions exist [1–6]. Sustainable agriculture as a term is defined by the American Society of Agronomy as "the one that, over the long term, enhances environmental quality and the resource base on which agriculture depends; provides for basic human food and fibre needs; is economically viable; and enhances the quality of life for farmers and the society as a whole." [7] (p. 15). According to the Sustainable Agriculture Initiative Platform, "sustainable agriculture is the efficient production of safe, high-quality agricultural products, in a way that protects and improves the natural environment, the social and economic conditions of farmers, their employees and local communities and safeguards the health and welfare of all farmed species." [8].

Sustainable agriculture is, therefore, the production of food and fibre in a way that the environmental, social, and economic dimensions are considered. The environmental dimension include issues such as climate change, energy, water scarcity, biodiversity, and soil degradation, while the social dimension is concerned with factors like labour rights, health of communities, accessibility and affordability of food, food quality and safety, as well as animal welfare. The economic dimension deals with productivity, efficiency, and competitiveness, where the benefits of these factors are not only noticed in farm profitability, but also in the rest of the value chain and will lead to thriving local economies [8].

According to Bongiovanni and Lowenberg-DeBoer [9], the concepts of precision agriculture (PA) and sustainability are inextricably linked. PA is a broad concept that has various definitions, but it can be described as a catch-all term for techniques, technologies, and management strategies that address in-field variability. It focusses on the development of integrated information and production systems that manage variability to optimise long-term, site-specific, and whole-farm productivity, and it also minimises the impact on the environment and natural resources. The application of fertilisers and water only where and when they are needed should reduce environmental loading, while the production is also more efficient and thus results in lower input costs per unit of output to enhance profitability. There exists a wide range of technologies that can be utilised to manage site-specific areas within a field. The adoption of these technologies is based on the farm scale, meaning that the level at which they become more cost-effective for a farmer depends on the cost savings for a farm, field, or different management zones multiplied by the area [10].

1.1. Problem Statement and Objectives

It is a simple task to calculate an enterprise budget and physical resource use for a certain crop under conventional farming (CF) practices, but it is more challenging to do so for different management zones and to calculate and evaluate the most sustainable practice for a particular farm or field. This is where the need arises to utilise a model that aids in the planning, analysis, and evaluation of two different scenarios for a particular farm and/or field. The large amount of variables, such as different crops, management practices, mechanisation technologies, variable rate irrigation (VRI), and variable rate applications (VRA), which must be considered for PA, highlight the need for a decision support model (DSM).

The main objective of the study is to investigate the impact of PA practices on the sustainability of a crop production enterprise, and the combination of enterprises as a whole-farm business, in comparison to CF. The sub-objectives are, firstly, to identify management zones according to variation in physical soil properties and,

91

secondly, to apply a DSM to evaluate the impact of PA practices on an individual farm enterprise and the farming business as a whole.

A multidisciplinary approach is needed for agricultural scenario planning, analysis, and evaluation of profitability and sustainability. There should be a combined focus on the following aspects, namely agricultural economics, agricultural mechanization, and agronomic principles. A farming operation is based on all of the above-mentioned aspects and the interaction between them, but in the end, the ultimate goal is to achieve financial stability and the sustainable use of natural resources.

2. Experimental Section

The study is based on irrigation fields situated on the western side of the Orange River in the Northern Cape Province of South Africa. The fields are situated on the 29° S latitude and 24° W longitude, at an altitude of 1024 m above sea level. The farm produces maize, soya beans, and wheat in a rotational system. Currently, a CF approach is followed where all the inputs (irrigation, fertilizer, and amelioration products) are uniformly applied over the entire field per crop. The input data that are used in the study were obtained from harvest monitor data, irrigation scheduling data, physical and chemical soil analysis, and historical data obtained from the farmer. The six fields that are used in the study cover a total area of 181.95 ha with an average of 30.32 ha per field.

2.1. Management Zone Identification

Management zones can be identified by using different approaches. The methods vary from soil type, soil texture, soil depth, precipitation, a combination of all these, and spatial variation in crop yield characteristics. Steven and Miller [11] suggest the use of multi-year yield maps. Accuracy and cost issues with the above-mentioned methods highlight the need for a remote sensing method to perform in situ measurements and a gamma-ray spectrometer (MS-1200 Type SBG932, Medusa Explorations: Groningen, Netherlands) was used in this study to take the measurements for management zone identification.

The correlations according to the count rate (Bq/kg) for the soil properties from the measurements obtained by the gamma-ray spectrometer are determined for clay, silt, and sand. The regression values that were respectively obtained for clay, silt, and sand were $R^2 = 0.979$, $R^2 = 0.810$, and $R^2 = 0.926$. The formulas obtained from the correlations are then used in a plant available water (PAW) model to extrapolate the specific property values to all the gamma-ray spectrometer readings. The PAW (mm) is calculated as:

$$PAW = FWC - WP \tag{1}$$

where *FWC* is the field water capacity (mm), and *W* is the wilting point in (mm).

After the calculation, interpolation and mapping of the PAW to the particular field boundaries with the use of Spatial Management Systems (SMS) software (*SMS Advanced*, 14.50; Ag Leader Technologies: Ames, IA, USA; 2014), the management zones for SMS can be defined. The physical and chemical properties of the soil can then be classified into the specific management zones. The VRI and VRA of fertiliser and amelioration products can then be planned in accordance to the crop yield potential of the particular management zone.

2.2. Decision Support Model

Due to the different management zones, the large number of variables that must be considered in calculating the enterprise budget for PA highlights the need to use a decision support model. The term decision support model (DSM) is broadly defined by Finlay as "a computer-based model supporting the decision-making process" [12] (p. 1282). The emphases of the DSM should be on supporting a certain decision with regard to a problem and not necessarily providing an answer. It must enable the farmer to base his/her decision on certain outcomes of different potential courses of action, thus, different scenarios. These scenarios can be based on economic, environmental, and social factors that may influence a specific choice or outcome.

The Scenario Planning, Analysis and Risk Evaluation (SPARÉ) model was designed in Microsoft Excel (*Microsoft Excel*, 2010; Microsoft: Redmond, WA, USA; 2010), to plan and evaluate two different scenarios under irrigation and/or dry land conditions with the use of multiple enterprise budgets per management zone and different crops per annual production cycle. There are certain designated sheets for the different production inputs, for instance fertiliser, lime and gypsum, mechanisation costs, chemical products, and water and electricity. These inputs can be changed per region, farm, season, etc., and the same cost is used for calculations in all scenarios [13].

The first step of the model is to use the different management zone areas and plan the farming operation accordingly. The initial farm planning consists of rotational crop planning per management zone per season, for irrigation and/or dry land according to a percentage of available area. After the initial planning is completed, individual crop enterprise planning should be done per management zone. This planning process consists of the following variables per zone: seeding, fertiliser, ameliorants, mechanisation, water demand and management, chemical products, and other costs. The following variable costs are taken into consideration to calculate and plan the whole farm business and each crop and management zone enterprise individually. The variable costs consist of seed, fertiliser, ameliorants, mechanisation, herbicides, pesticides, insurance, irrigation, transport, marketing, other variable costs, and interest on operating capital. All of these costs are taken into account to calculate the impact on each enterprise in accordance to the whole farm operation.

The model's structure is described in Figure 1, which gives an overview of the model as a whole from farm information, management zone planning, enterprise planning, and enterprise budgets to farm income summary, evaluation, and analysis.

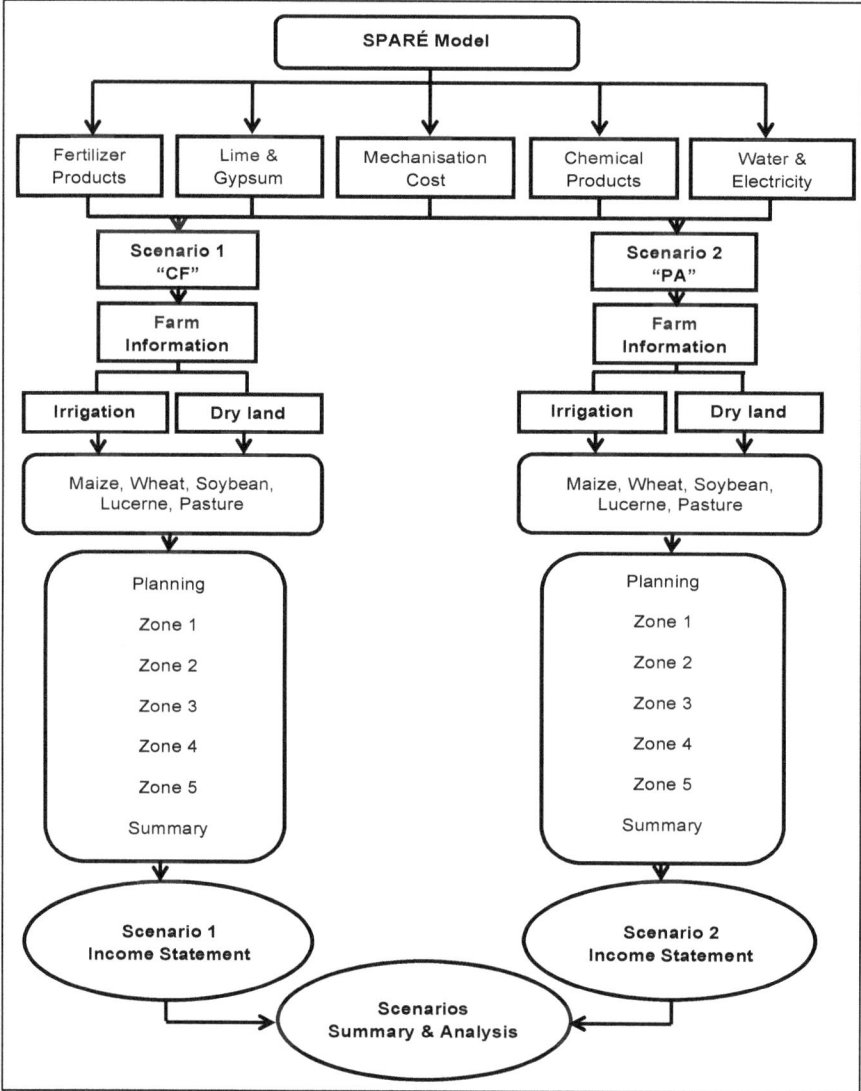

Figure 1. Schematic representation of the SPARÉ model.

All calculations start from management zone level, then the enterprise level to whole-farm level (all calculations and formulas are available from the authors upon request). The gross margin (GM) of the scenario (SC) is the final answer with regard to profitability and is calculated as:

$$GM_{SC} = GI_{SC} - TVC_{SC} \tag{2}$$

where GI_{SC} is the gross income (GI) of the scenario, and TVC_{SC} is the total variable cost (TVC) of the scenario.

The total income of all the management zones gives the sum of the specific enterprise and the total of the enterprises gives the sum for the specific scenario. The GI of a SC is calculated as:

$$GI_{SC} = GI_{E(I_a + I_b + I_c + I_d + D_a + D_b + D_c + D_d)} \tag{3}$$

where E is enterprise; $I_\#$ is irrigation enterprise; and $D_\#$ is dry land enterprise.

The cost calculations consist of variable costs and are the part of the total cost component that could vary within the framework of a specific production structure, as the size of the enterprise varies and/or the intensity of the production per unit changes. The TVC of a SC is calculated as:

$$TVC_{SC} = TVC_{E(I_a + I_b + I_c + I_d + D_a + D_b + D_c + D_d)} \tag{4}$$

Financial analysis pertains not only to income and expenditure, but also to the ability to meet financial liabilities, carry risk, and strategically utilise available capital. The break-even price and yield are simple calculations that can be used to calculate the minimum price and yield that must be achieved for a particular management zone or enterprise to be profitable. The operating profit margin ratio is used to measure the operating efficiency of a farm business and it is usually expressed as a percentage. The operating profit margin (OPM) for an enterprise E is calculated as:

$$OPM = \left(-\frac{GM_E}{GI_E} \right) \% \tag{5}$$

The SPARÉ model also calculates the total amount of variable inputs used for the different management zones, enterprises, and the whole-farm operation in physical quantities. This information can then be used to determine the change in input use efficiency between PA and CF.

3. Results and Discussion

3.1. Identified Management Zones

The PAW (mm) of the fields is shown in Figure 2. The field's clay percentages vary between 5% and 30% and the PAW varies between 35 mm to above 50 mm. The infiltration rate is directly correlated with the clay percentage and it varies between 25 mm· hr^{-1} to as low as 8 mm·hr^{-1}. From the variation in spatial PAW data, five management zones in pie slice-shaped sectors are identified. The management zones (sectors) differ in segments of five from below 35 mm to above 50 mm. The zones are, respectively, 13.9, 47.8, 47.1, 57.3, and 15.6 ha.

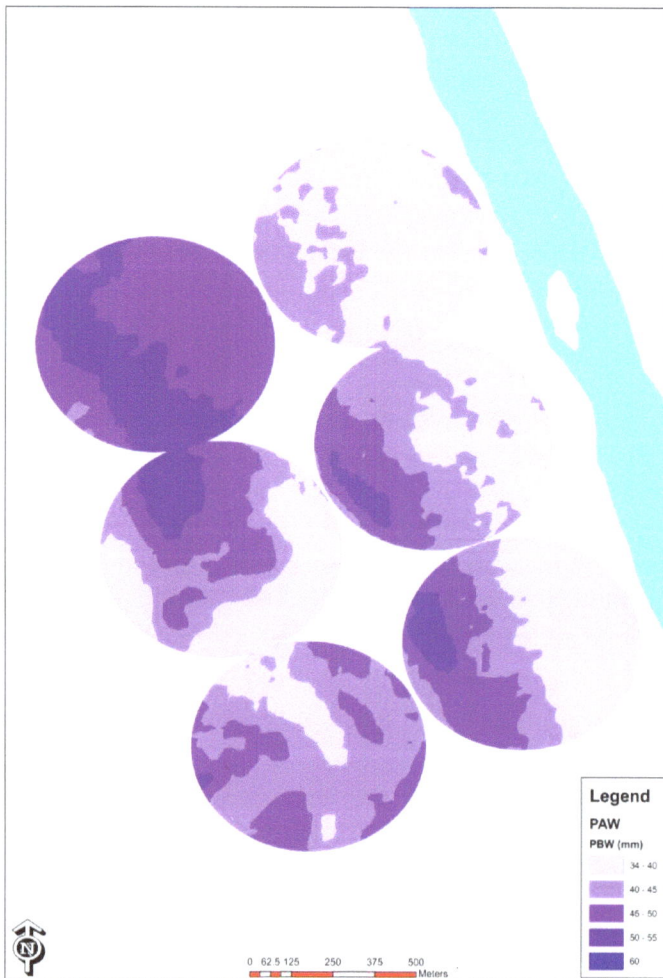

Figure 2. Plant available water map of the study fields.

These identified management zones are used in the decision support model for the PA calculations. The VRI and VRA of fertiliser and amelioration products are then applied in the model in accordance to the crop yield potential of the particular management zone.

3.2. Impact of Precision Agriculture

The impact of PA on the sustainability of crop production will first be discussed on the basis of economic sustainability and then on the basis of environmental sustainability. Figure 3 presents a summary of the total income, total variable cost, and the gross margin for CF and PA. It is evident from Figure 3 that although the total variable cost for PA is higher than that for CF, the much higher total income from PA results in a higher gross margin. The total variable cost increase of 0.7% for PA is, thus, offset by the 10% increase in total income and results in an increase of 26.9% in the gross margin.

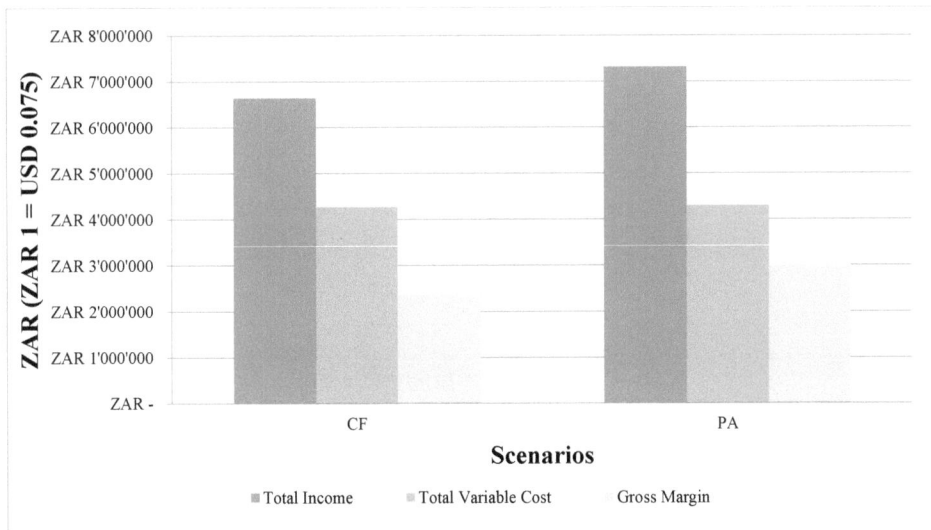

Figure 3. Summary of the total income, total variable cost, and gross margin of conventional farming (CF) and precision agriculture (PA) in South African Rand (ZAR).

The operating profit margin (OPM) is 36% and 41%, respectively, for CF and PA for the whole-farm scenario. Thus, it is 5% higher in the case of PA, making PA more profitable than CF. This also means that PA has a higher return on investment (ROI) than CF for each South African Rand (ZAR) spent.

When comparing the individual enterprises according to CF and PA, it is evident that PA is more profitable than CF. The GM for maize, wheat, and soya beans are, respectively, 22.3%, 27%, and 36.2% higher for PA than for CF. The OPM of CF and PA for maize is 32% and 37%, respectively, for wheat it is 48% and 54%, respectively, and for soya beans it is 20% and 27%, respectively. From these figures it is evident that PA practices are more profitable than CF, with the correct ratio of in-field variation.

While the use of PA does not mean that the total amount of resource use will be less than with CF, by managing the in-field variability the resource use is usually more effective. In order to determine the influence of PA on the environmental sustainability of crop farming it is, thus, necessary to compare the resource use of PA and CF with the yield. The physical total quantities of resources (for example fertiliser, lime, gypsum, fuel, and water) used in tons, kilograms, or litres, are thus divided by the total tonnage of yield for the enterprise to calculate the resource use per ton of output.

The differences in variable input use for the whole-farm scenario between PA and CF are presented in Figure 4. The variable input with the largest saving is the amount of irrigation water applied, which is 24.1% lower in the case of PA. The variable rate irrigation system that is used for PA only applies water where needed and, therefore, the total water use in the case of PA is only 180 mm·ton^{-1}, while it is 237 mm·ton^{-1} for CF. A total amount of water of 57 mm·ton^{-1} of yield is, thus, saved for the specific fields when PA is used instead of CF. The quantities of lime and gypsum, fertiliser and amelioration fertiliser that are used for PA, also differ significantly from CF. In the case of PA, 22.6% less lime and gypsum was applied than with CF, which was a total of 58 kg·ton^{-1} of yield. The amelioration fertiliser use was 2 kg·ton^{-1} of yield less for PA, while fertiliser use was 24 kg·ton^{-1} of yield less.

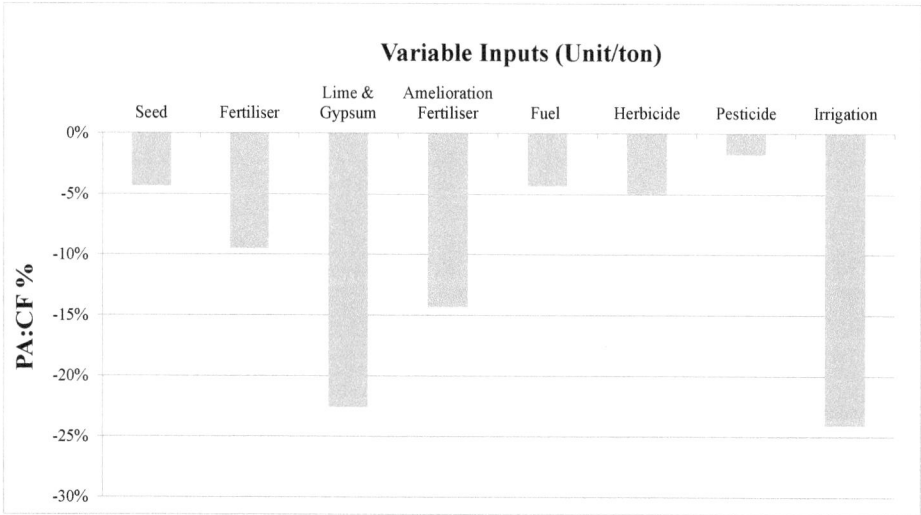

Figure 4. Differences between the variable input use of PA and CF.

4. Conclusions

The findings of this article prove that PA may have a positive impact on the sustainability of crop farming for both the economic and environmental dimensions of sustainability. The more sustainable economic dimension was proved by the higher gross margin realised by PA, while the environmental dimension is more sustainable due to the more efficient use of variable inputs, such as water and fertiliser. This finding confirms the findings of Sadler, Evans, Stone, and Camp [14], who found that variable rate irrigation can reduce water usage, and of Lencsés, Takács, and Takács-György [15], who state that PA can reduce the harmful effects of chemical use. The positive impact on the economic dimension will also lead to a more sustainable social dimension of sustainability, as PA produces food more efficiently and, thus, enhances both the availability and affordability of food.

It must, however, be kept in mind that the feasibility of PA practices depends on in-field variation, crop value, economies of scale and the useful life of the equipment. According to Maine "PA has the potential to enhance profitability on South African soils, which are characterised by great variability in depth and fertility within given fields" [16]. VRI is projected to become more essential in the future to protect the scarce water resources in South Africa and the world. Efficiency in agriculture will also become more significant in the future, as more food must be provided to a growing population by using a limited amount of natural resources.

PA is certainly not a new concept, but the adoption of this technology occurs at a relatively slow rate. The higher investment cost in the case of PA (as opposed to CF) and the difficulty to calculate the possible returns beforehand withhold producers from adopting it. This article not only shows what the economic and ecologic advantages of PA can be, but also provides a model that can be applied to calculate these advantages for the producer who considers PA as an alternative to CF.

Author Contributions: Frikkie Alberts Maré was responsible for the writing of this article and assisted with the development of the SPARÉ model. Hendrik Petrus Maré identified the management zones with the gamma-ray spectrometer, assisted in the development of the SPARÉ model and collected data for the model from the farmers. Both authors have read and approved the final manuscript.

Conflicts of Interest: The authors declare no conflict of interest.

References

1. Solow, R. The Economics of resources or the resources of economics. *Am. Econ. Rev.* **1974**, *64*, 1–13.
2. Hartwick, J. Substitution among exhaustible resources and intergenerational equity. *Rev. Econ. Stud.* **1978**, *45*, 347–354. [CrossRef]
3. WCED (World Commission on Environment and Development). *Food 2000: Global Policies for Sustainable Agriculture*; Zed Books: London, UK, 1987.

4. Pearce, D.; Atkinson, G. Capital theory and the measurement of sustainable development. As Indicator of Weak Sustainability. *Ecol. Econ.* **1993**, *8*, 103–108. [CrossRef]

5. Pearce, D.; Atkinson, G. Measuring of Sustainable Development. In *The Handbook of Environmental Economics*; Bromley, D., Ed.; Wiley-Blackwell: Oxford, UK, 1995; pp. 166–181.

6. Caffey, R.H.; Kazmierczak, R.F.; Avault, J.W. Incorporating Multiple Stakeholder Goals into the Development and Use of a Sustainable Index: Consensus Indicators of Aquaculture Sustainability. *Dep. Agric. Econ. Agribus. La. State Univ. USA. Staff Pap.* **2001**, *8*, 40.

7. American Society of Agronomy. Decision reached on sustainable ag. *Agronomy News*, January 1989; 15.

8. Sustainable Agriculture Initiative Platform. Available online: http://www.saiplatform. org/sustainable-agriculture/definition (accessed on 02 September 2015).

9. Bongiovanni, R.; Lowenberg-DeBoer, J. Precision Agriculture and Sustainability. *Precis. Agric.* **2004**, *5*, 359–387. [CrossRef]

10. Bootle, B.W. Precision agriculture. Final Report to the Australian Nuffield Farming Scholars Association. Australian Nuffield Farming Scholars Association: Ningan, NSW, Australia, 2001.

11. Steven, M.D.; Millar, C. Satellite Monitoring for Precision Farm Decision Support. In *Precision Agriculture '97. Proceedings of the First European Conference on Precision Agriculture, Warwick University, Oxford, UK, 7–10 September 1997*; Stafford, John V., Ed.; BIOS Scientific Pub.: Oxford, UK, 1997; pp. 697–704.

12. Finlay, P.N. *Introducing Decision Support Systems*; Blackwell: Oxford, UK, 1994.

13. Maré, H.P.; Maré, F.A. A decision support model for the adoption of precision agriculture practices. In Proceedings of the 20th International Farm Management Congress, 12–17 July 2015; Laval University: Québec City, Québec, Canada, 2015; pp. 268–280.

14. Sadler, E.J.; Evans, R.G.; Stone, K.C.; Camp, C.R. Opportunities for conservation with precision irrigation. *J. Soil Water Conserv.* **2005**, *60*, 371–379.

15. Lencsés, E.; Takács, I.; Takács-György, K. Farmers' perception of precision farming technology among Hungarian farmers. *Sustainability* **2014**, *6*, 8452–8465. [CrossRef]

16. Maine, N. The profitability of Precision Agriculture in the Bothaville District. Ph.D. Thesis, University of the Free State, Bloemfontein, South Africa, 2006.

Factors Affecting Sheep Theft in the Free State Province of South Africa

Willem Lombard, Walter van Niekerk, Antonie Geyer and Henry Jordaan

Abstract: Livestock theft has a big impact on the livestock industry of South Africa and is threatening the sustainability of the industry. In order to generate information that can be used to inform sheep farmers on how to effectively mitigate stock theft on their farms, the objective of this study was to investigate whether the factors affecting the occurrence of livestock theft are significantly different from the factors affecting the level of livestock theft experienced in the Free State Province of South Africa. The study was based on data captured in 292 structured questionnaires completed during telephonic interviews with livestock farmers in the Free State Province. The Craggs model specification was used to statistically test whether or not the same factors affect the occurrence of stock theft and the level of stock theft experienced by the respondents. The results revealed that factors affecting the occurrence of stock theft are significantly different from the factors affecting the level of livestock theft. Technologies used by farmers proved to be significantly related to the occurrence of livestock theft, while loss controlling actions taken by farmers proved to have significant relationships to the level of livestock theft experienced. Technologies used include stock theft collars and alarms. Loss controlling actions include night patrols, counting animals and access control. The results proved that the stock theft problem faced by farmers can be divided into occurrence and the level of occurrence aspects. Thus, investing in controlling actions may decrease the level of stock theft, but not necessarily stop stock theft. Other challenges faced by farmers that threaten the sustainability of their farming enterprises should be approached in a similar manner to generate information that can be used to more effectively overcome the challenge at hand.

---

1. Introduction and Background

Livestock theft is nothing new to South African farmers and is considered by some to be as old as farming itself [1,2]. Recorded cases of livestock theft in South Africa can be traced back to 1806 [3]. In some African cultures, cattle raiding (livestock theft) formed a major part of warfare. It was even considered legitimate to enter neighbouring chiefdoms and raid their cattle during times of peace. These raiders who returned with large numbers of cattle were seen as heroes, while petty thieves were despised [4]. Livestock theft is not a problem that is unique to South Africa or even Africa. Various countries also experience livestock theft and have done research to try and identify causes and solutions to this problem. African countries include

101

Lesotho [5], Kenya [6–8] Eritrea [9] and Nigeria [10], while other countries include the USA [11] and Australia [12]. From reviewed literature, it seems that livestock theft has become more violent and organized in recent years, e.g., with guns are used in perpetrating these thefts. One of the main causes of livestock theft is poverty among unemployed and drought-stricken crop farmers [5,7,13].

Livestock theft statistics show that all nine South African provinces are victims of stock theft [14]. The annual economic impact of livestock theft on the South African red meat industry (sheep, cattle and goats) for the year 2011/12 was estimated at 300 million South African Rand (ZAR) [15]. This amount is far less than the estimate of Clack [2] who calculated the 2011/12 annual losses at approximately ZAR 487 million. The total cost of losses to the red meat sector further increased to approximately ZAR 514 million in 2013/14 [16].

It should be noted that farmers not only have to deal with controlling livestock theft [2] but also other problems such as predators [17,18] and extreme weather conditions (draught, animal diseases, etc.) [19]. As the cost of controlling these increasing problems, more pressure is placed on the farmer's profit margin. In some cases, livestock farmers have already left the livestock industry because of stock theft, resulting in a shortage of supply and increased prices, threatening sustainability [1].

1.1. Problem Statement

Despite the significant losses associated with livestock theft in South Africa, the topic has received very little attention from researchers. The authors of this study could not find any research of factors affecting the occurrence of livestock theft under South African conditions. Thus, no scientific evidence is available to advise farmers on how to control livestock theft. The aim of this study was to explore the factors that have an effect on sheep theft in the Free State Province. The factors were analysed to determine whether factors affecting the occurrence of livestock theft are the same as factors affecting the level of livestock theft experienced. As such, a better understanding of the problem faced by the producers could contribute more effective response strategies to mitigate livestock theft in South Africa.

1.2. Study Area

The Free State Province of South Africa, which is the focus of this study, is situated centrally within South African borders (Figure 1). The Free State Province is divided into five district municipalities: Fezile Dabi, Lejweleputswa, Mangaung, Thabo Mofutsanyane and Xhariep. The province does not only share its border with six other provinces, but also with Lesotho. Lesotho, also known as the Mountain Kingdom, is completely surrounded by South Africa [20]. The border shared between the Free State Province and Lesotho is 450 km long and is guarded by 100 troops of the South African National Defence Force (SANDF) [21]. The Free State Province has a population of 2,745,590 [22] with roughly 54,000 people employed in the agricultural sector of the province [23]. According to the Department of Agriculture, Forestry and Fisheries (DAFF) there are 6065 commercial livestock farming units in the Free State Province [24]. The province has a total size of 12,943,700 ha, of which 90.9% is used for farming [24]. Grazing land, which is mainly suitable for livestock farming, makes up 58.1% of commercial farmland and 66% of emerging farmland [24]. The Free State Province has the third largest number of sheep as well as cattle, estimated at approximately 4.8 million sheep and 2.3 million cattle respectively [25].

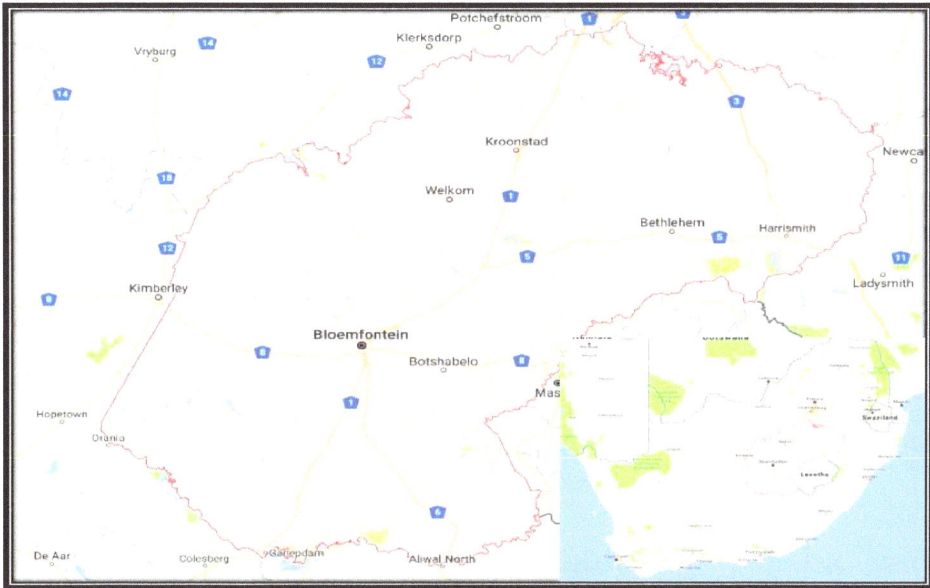

Figure 1. Geographical location of the Free State Province [26].

1.3. Data

A structured questionnaire was developed to obtain relevant information regarding livestock theft in the Free State Province. The questionnaire was designed to be administered during telephonic interviews. The questionnaire was designed based on the principals suggested by [27] and on a questionnaire used by van Niekerk [17]. The questionnaire included questions on farmers' years of farming experience, age, farm size, farm location and farm topography, losses due to livestock theft and practices used to control livestock theft. Questions of the practices used to control livestock theft included methods used, actions taken, how often these practices are performed and the annual cost of these practices.

Stratified random sampling was applied to select the respondents. Stratified random sampling is where the population is divided into subpopulations and random samples are then chosen from each subpopulation [28,29]. Livestock farmers within the Free State Province were divided into different subpopulations within the province according to their farm's demographic and topographic location. This method of sampling was chosen so that comparison and correlation between the different subpopulations can be done. By following this method, it also ensured that only livestock farmers were interviewed. The interviews took place between May and August 2014. Most of the farmers were contacted during the late afternoon and early evening. In total, 292 farmers completed the questionnaires and were included in further analyses.

2. Experimental Section

The main objective of this study was to identify factors affecting livestock theft in the Free State Province. Van Niekerk [17] found that the factors affecting the occurrence of predation and the factors affecting the level of predation are not the same. Based on this, it was hypothesised that the factors affecting the occurrence of livestock theft and factors affecting the level of livestock theft are not the same. In this study, the model used to investigate factors affecting livestock theft cover two aspects: whether or not livestock theft occurred and if livestock theft did occur, what quantity (level or the number of animals) of livestock were stolen. Similar to van Niekerk [17], the Craggs model was used to scientifically test whether the factors that affect the occurrence of livestock theft are the same as the factors that affect the level of livestock theft.

The Craggs model allows for one set of parameters to determine the probability of livestock theft occurring and another set of parameters to determine the number of livestock stolen (level of livestock theft) [30,31]. Due to the fact that it is hypothesised that the factors affecting the occurrence of livestock theft and factors affecting the quantity of livestock theft are not the same, the Craggs model would be an appropriate model for the study.

The Probit model was used to model whether or not livestock theft occurred (yes/no) and the Truncated model was used to measure the level (how much) of livestock theft experienced. According to Katchova and Miranda [32] the Probit (1) and Truncated (2) models are represented as follows:

Probit:

$$P(\alpha_i = 0) = \Phi(-\frac{\beta'_\alpha X_i}{\sigma})$$

(1)

where P = is the probability, α_i = quantity of livestock theft, Φ (.) = standard normal probability density function, β'_α = a vector of coefficients, X_i = variable or an $S \times 1$ vector of personal and farm characteristics for farmer i and σ = variance.

Truncated:

$$f(\alpha_i | \alpha_i > 0) = \frac{f(\alpha_i)}{P(\alpha_i > 0)} = \frac{\frac{1}{\sigma}\Phi\left(\frac{\alpha_i - \beta'_\alpha X_i}{\sigma}\right)}{\Phi\left(\frac{\beta'_\alpha X_i}{\sigma}\right)} \tag{2}$$

where: $f(.)$ = the probability density function, P = the probability, α_i = the density (quantity) for the positive values, $\Phi(.)$ = standard normal probability density function, β'_α = a vector of coefficients, X_i = a variable or a $S \times 1$ vector of personal and farm characteristics for farmer i and σ = variance.

It is important to note that the Tobit model returns when the occurrence of livestock theft estimated in the Probit model (1) and the level of livestock theft experienced modelled in the Truncated model (2) have the same factors X_i and the same parameter vector β'_α [32]. Lin and Schmidt [30] prescribe the Lagrange multiplier to test the restrictions of the Tobit model. Greene, Woodruff and Tueller [33] suggests that the restrictions could be tested by calculating the following log-likelihood test statistic (3) after the Truncated model, the Tobit model and Probit model have been calculated.

$$\lambda = -2\left[\ln L_{Tobit} - \left(\ln L_{Probit} + \ln L_{Truncated\ regression}\right)\right] \tag{3}$$

where: λ = likelihood ratio statistic, L_{Tobit} = likelihood for the Tobit model, L_{Probit} = likelihood for the Probit model and $L_{Truncated\ regression}$ = likelihood for the Truncated model.

If the Cragg model has a significant p-value (probability), the factors affecting livestock theft differ significantly from the factors affecting the quantity of livestock theft. If, however, an insignificant p-value is found, the factors affecting the occurrence and quantity of livestock theft are the same and the Tobit model should be efficient for the analysis. The Cragg test was conducted in NLOGIT 4.0 statistical software (*NLOGIT*, 4.0; Econometric Software, Inc.: New York, NY, USA, 2006).

Hypothesised Factors

In an investigation of the available literature on livestock theft and the control of livestock theft, different factors were identified that affected the occurrence of livestock theft. Two main types of factors can be identified, namely internal and external factors. These main types of factors can further be divided into sub-groups. External factors include factors that the farmer has little or no control over. Identified external factors can be divided into demographic factors and topographic factors. Demographic factors include factors such as the age of farmers and topographic factors include farm size and distance from town. Internal factors are the factors that a farmer can control and include management practices for stock theft prevention and detection; physical barriers for stock theft prevention and detection; technological systems for stock theft prevention and detection; animals used for stock theft prevention and detection; and livestock insurance in South Africa. All factors considered in this study were placed in one of the sub groups.

Note that the hypothesised factors were tested for multicollinearity. None of the factors proved to have a variance inflation factor above the cut-off value of 10, therefore no multicollinearity was found to be present.

3. Results and Discussion

The factors affecting livestock theft were investigated by means of a Probit, Tobit and Truncated regression model specification. The external and internal factors were analysed separately and are discussed as such.

3.1. External Factors Affecting Livestock Theft

Results for the external factors hypothesised to affect the occurrence and level of livestock theft in the Free State Province consist of Tobit (level), Probit (occurrence) and Truncated (level) results which are shown in Table 1. The Craggs test was used to determine whether the variables affecting the occurrence of livestock theft are significantly different from the factors affecting the level of livestock theft experienced. If the factors prove to affect both the occurrence and level of stock theft experienced, the Tobit model would have been the model of choice. However, if the factors affecting the occurrence of stock theft prove to be different form the factors affecting the level, the Probit and Truncated model specifications have to be used to separately model the probability of occurrence and the level of occurrence.

Table 1. Regression results of the Tobit, Probit and Truncated specifications when analysing external factors affecting livestock theft.

Variable	Tobit	Probit	Truncated
Dependent variable	Number of Sheep Stolen	Dummy = 1 if Experienced Theft, Otherwise 0	Number of Sheep Stolen
Constant	−173.4253 ****	−3.2335 ****	85.2656
	(56.3445)	(0.8090)	(269.6226)
Reporting of livestock theft			
Report within 0 and 1.99 h	108.4810 ****	1.7436 ****	−17.8845
	(18.5546)	(0.2446)	(88.3507)
Report within 2.00 and 4.99 h	135.0665 ****	1.7156 ****	104.1057
	(24.3078)	(0.3389)	(101.9826)
Report within 5.00 and 12.99 h	113.0441 ****	2.1386 ****	−66.6959
	(34.5546)	(0.5952)	(135.7859)
Report within 13.00 and 24.00 h	110.4856 ****	0.9739 **	226.0078 **
	(38.5791)	(0.5293)	(131.2323)

Table 1. *Cont.*

	Tobit	Probit	Truncated
Management of farm workers			
Average relationship with herdsman	38.8593	0.2611	54.8602
	(32.9979)	(0.5011)	(122.4384)
Good relationship with herdsman	41.2272	0.5193	27.1038
	(28.7994)	(0.4468)	(113.2347)
Very good relationship with herdsman	16.5898	0.2078	−36.9059
	(28.8577)	(0.4435)	(115.4441)
Take ID copy	32.0676	0.0026	119.3469
	(32.3090)	(0.5289)	(100.5505)
Check employees' history	−17.0609	−0.0676	−123.6627**
	(21.3817)	(0.3485)	(63.6703)
Pay workers on weekly basis	−29.3835	−0.3287	−35.0568
	(50.0152)	(0.7251)	(218.5194)
	Tobit	**Probit**	**Truncated**
Management of farm workers			
Pay workers on monthly basis	36.9110	0.9789 *	51.6999
	(44.9871)	(0.6690)	(221.97320
Workers go to town every weekend	3.5809	0.0426	−10.5054
	(27.6891)	(0.4206)	(102.6302)
Workers go to town every second weekend	−37.9533	−0.3076	−264.2264 **
	(31.1944)	(0.4720)	(143.3688)
Workers go to town once a month	−30.4430	−0.3608	−99.3744
	(25.7465)	(0.3913)	(96.5304)
Workers receive visitors	24.9691	0.1464	91.0263
	(24.4536)	(90.3518)	(97.9315)
Visitors walk through farm	−0.3027	0.1936	−45.3874
	(14.0565)	(90.2237)	(47.7825)
Number of employees	−0.3118	−0.0126)	0.8742
	(1.0411)	(90.0157)	(3.7305)
Demographic factors			
Years farming	0.2528	0.0028	1.0222
	(0.4992)	(0.0077)	(1.7542)
Age	−0.1515	0.0017	−1.0781
	(0.5561)	(0.0083)	(2.1080)
Fulltime farmer	−15.7624	0.0501	−105.1946*
	(19.3036)	(0.2932)	(67.2876)
Topographic factors			
Plains	15.4593	0.8169 ****	−176.8991 ***
	(19.2827)	(0.2909)	(72.4865)
Mountains	16.1734	0.0192	73.6144
	(16.3314)	(0.2667)	(56.3414)
	Tobit	**Probit**	**Truncated**
Topographic factors			
Planted pastures	−8.6230	−0.2158	25.0973
	(12.4697)	(0.1932)	(45.4221)
Distance from town	1.3980	0.0392 **	−4.2555
	(1.4501)	(0.0211)	(4.7594)
Distance to informal settlement	−1.1857	−0.0279	2.4076
	(1.4783)	(0.0217)	(4.7692)

Table 1. *Cont.*

	Tobit	Probit	Truncated
Size of farm	0.0070 ***	0.0052	0.0195 ***
	(0.0031)	(0.0049)	(0.0097)
Large town	−7.8256	−0.2227	29.8487
	(17.4066)	(0.2533)	(63.8693)
Border	45.3178 **	1.4644 ****	27.4206
	(24.9301)	(0.5467)	(80.6962)
Stock theft hotspot	9.1803	0.1651	28.8618
	(20.0244)	(0.3010)	(73.5946)
	Tobit	Probit	Truncated

Goodness of Fit			
No. of observations	292	292	292
Sigma	84.0267 ****	—	113.0711 ****
	(4.8356)	—	(12.0814)
Log likelihood	−1034.1803	−199.6506	−830.5208
% Correct prediction	—	77.055%	—
McFadden R^{2a}	—	0.2933	—
Model chi-square [b]	—	117.0961	—
Significance level [c]	—	(0.0000)	—
Likelihood-ratio test for Tobit vs. Truncated regression[d]	—	—	125.1138[d]
	—	—	(0.0000)[c]

****, **, * denote 1%, 5% and 15% significance level respectively. Standard errors are in parentheses. [a] McFadden R^2 is given by 1 − the ratio unrestricted:restricted log likelihood function values. [b] The chi-square test evaluates the null hypothesis that all coefficients (not including the constant) are jointly zero. [c] Numbers in parentheses are associated with chi-square probabilities. [d] The likelihood ratio test is given by $\lambda = 2 (\ln L_{Probit} + \ln L_{Truncated\ regression} - \ln L_{Tobit})$.

The aim of these regressions was not to predict the probability of livestock theft but rather identify the internal and external factors associated with a lower probability of livestock theft. Thus, a significance level of 15% was considered acceptable for indicating a statistically significant relationship. In order to ease discussion and identify trends, external factors were divided into suitable categories: reporting of livestock theft, management of farm workers, demographic factors and topographic factors.

The Graggs test indicated whether the factors affecting the occurrence of livestock theft are significantly different from the factors affecting the level of livestock. Results from the Graggs test (Table 1) indicate that the log-likelihood test ratio of 125.11 is highly significant ($p < 0.01$). Therefore, the Tobit specifications are rejected in favour of the more general Graggs model specification. Thus, external factors affecting the occurrence of livestock theft are significantly different from the factors affecting the level of livestock theft in the Free State Province.

The Probit regression (Table 1) identified eight external factors that have a significant relationship with the occurrence of livestock theft in the Free State Province. The reporting of livestock theft shows that all of the reporting options offered to farmers proved to be significant: "report within 0 and 1.99 hours" ($p < 0.01$), "report within 2.00 and 4.99 hours" ($p < 0.01$), "report within 5.00 and 12.99 hours" ($p < 0.01$), "report within 13.00 and 24.00 hours" ($p < 0.10$). Strangely, all of these factors proved to be positively related to the occurrence of livestock theft. Thus, it does not matter how long it took to report the theft, the probability of the occurrence of livestock theft increases. Under the management of the farm worker factors, "paying workers on a monthly basis" ($p < 0.15$) showed a positive relationship to the occurrence of livestock theft. This implies that where farm workers were paid once a month, there was a higher probability for the occurrence of livestock theft. Results for the topographic factors showed that "plains" ($p < 0.01$), "distance from town" ($p < 0.10$) and "border" ($p < 0.01$) all related with the occurrence of livestock theft in a positive direction. Thus, farms with more plains (flatter land) are more likely to experience livestock theft. This contradicts the findings of Barclay and Donnermeyer [12] who found that higher stock theft rates are experienced in hilly terrain. It should also be taken into account that large parts of the Free State Province are relatively flat. Farms further away from towns have a higher probability for the occurrence of livestock theft; this agrees with the findings of Barclay and Donnermeyer [12] but contradict the findings of Bunei et al. [8]. One could argue that isolated farms create the opportunity for theft without being seen by the farmer. Lastly, farms close to the Lesotho border are more likely to experience livestock theft.

The external factors that have a significant relationship with the level of livestock theft experienced in the Free State Province are shown by the Truncated results in Table 1. Contrary to the result from the Probit model, only the "reporting theft within 13.00 and 24.00 hours" ($p < 0.10$) variable proved to be significant in the reporting of livestock theft category. Thus, farmers who experienced a higher level of stock theft tended to report a crime 13.00–24.00 h after it was committed. Management of farm workers had two significant factors: "checking employees' history ($p < 0.10$) and "workers go to town every second weekend" ($p < 0.10$). Both of these factors had a negative sign for their coefficient. The results thus suggest that checking employees' history and taking workers to town every second weekend are associated with lower levels of livestock theft. One reason why checking employees' history was associated with lower levels of livestock theft could be that no farmer would hire a known criminal. In cases where workers are taken to town every second weekend it could ensure that they are able to buy enough food in town so that they do not need to steal livestock for food, if that were the case. It could also be a sign that farm workers are involved in organised crime and could be serving as informants to criminals when not taken to town, however this result should not be generalized for all farmworkers.

Demographic factors indicated that "fulltime farmers" ($p < 0.15$) experience a lower level of stock theft. This could be due to the fact that fulltime farmers usually have more time to check up on the livestock and can detect any strange activity on the farm during the day. Topographic factors that proved to be significant with the level of stock theft experienced was "plains" ($p < 0.05$) and "size of the farm"($p < 0.05$). When interpreting the direction of signs, plains had a negative sign, meaning farms which have more plains (flatter) experience a lower level of livestock theft. It could be that a thief will not easily be able to hide a large number of animals in a flat area but it could easily be done in mountainous terrain. The size of the farm had a positive relationship, which means that farmers who have larger farms have experienced higher levels of livestock theft. This is simply because a farmer will struggle to focus on the whole farm simultaneously. Paddocks far from the farm house might also not be in the line of sight to detect any strange activity immediately.

3.2. Internal Factors That Affect Livestock Theft

The internal factors hypothesised to affect the occurrence and level of livestock theft in the Free State Province were analysed and the results are shown in Table 2. In order to ease discussion and help identify trends, external factors were also divided into groups: management practices, physical barriers, technology used, animals used and actions taken against stock theft.

Table 2. Regression results for the Tobit, Probit and Truncated specifications when analysing internal factors influencing livestock theft.

Variable	Tobit	Probit	Truncated
Dependent Variable	Number of Sheep Stolen	Dummy = 1 if Experienced Theft, Otherwise 0	Number of Sheep Stolen
Constant	−65.1119 ****	−0.6914 **	−210.2690 **
	(25.0723)	(0.3668)	(111.7880)
Management practices			
Guards	34.5417 **	0.4824 *	25.6400
	(17.8570)	(0.2983)	(54.5760)
Strategic Guard	−15.3475	0.2301	−200.0290
	(32.9571)	(0.5915)	(161.2492)
Theft informant	167.7423 ****	1.0776	340.2137 ****
	(39.2791)	(0.7645)	(86.7694)
Strategic Theft informant	29.0855	0.3976	129.3073 **
	(24.5194)	(0.4269)	(75.0451)
Physical barriers			
Corral at night	42.6286 ****	0.9725 ****	−34.0825
	(12.9723)	(0.2045)	(48.4926)
Strategic Corralling	26.4890 *	0.1120	168.5605 ****
	(17.4323)	(0.2827)	(64.3844)
Lock gates	32.8182	0.6940	−210.6150
	(54.71061)	(0.8656)	(378.8318)
Electric fencing	−0.6457	−0.2567	140.8787
	(30.5192)	(0.4543)	(107.4632)
Strategic Electric fences	−73.6186	−1.1324	−37.4893
	(58.3410)	(0.7962)	(331.9862)

Table 2. *Cont.*

Variable	Tobit	Probit	Truncated
Technology used			
Stock theft collars	55.6696 ****	0.9733 ****	128.1724 ***
	(20.1175)	(0.3713)	(63.1425)
Lights in corral	−59.0573	−1.2619 *	172.9464
	(58.8714)	(0.8661)	(271.1620)
Alarm in corral	24.4059	1.2170 ***	−134.6190
	(30.1628)	(0.5705)	(120.4728)
Camera	80.7218 ****	0.1899	236.6964 ****
	(23.0408)	(0.3820)	(66.2950)
Strategic Stock theft collars	50.8349 **	1.2312 ***	1.1539
	(30.4859)	(0.6117)	(86.6459)
Strategic Camera	37.8077	0.6102	−45.2779
	(35.2207)	(0.6110)	(109.6997)
Animals used			
Ostrich	35.9263	0.4373	156.4953
	(44.8285)	(0.6499)	(134.5210)
Donkey	−4.3010	0.0990	−136.4890
	(27.6415)	(0.4161)	(121.2614)
Wildebeest	−41.6752	0.1438	−69.1198
	(83.2292)	(1.1605)	(236.7963)
Animals used			
Dogs	8.1136	0.0788	50.9674
	(22.5095)	(0.3612)	(89.4025)
Strategic Dogs	−15.2595	0.3690	−123.5780
	(30.0255)	(0.5280)	(111.2376)
Actions taken against stock theft			
Active patrols	32.9270 ***	0.1381	156.1180 ****
	(13.5464)	(0.2002)	(57.9760)
Access control	−1.4453	0.2898	−108.1990 *
	(15.3461)	(0.2303)	(67.5339)
Strategic Patrols	14.2459	0.2953	5.5086
	(18.1697)	(0.2869)	(64.7334)
Strategic access control	29.9737 **	0.0083	103.5554 **
	(17.7402)	(0.2829)	(54.7067)
Count daily	−18.1366	−0.3721 *	−35.1218
	(15.0682)	(0.2302)	(53.3414)
Count more than once per day	−34.7825	−0.6842	−40.3194
	(42.0292)	(0.5651)	(215.4136)
Count once per week	−0.7589	0.0944	−78.5614
	(15.0015)	(0.2269)	(57.0154)
Count more than once per week	24.0160	0.5281 ***	−9.9554
	(17.2187)	(0.2691)	(57.2689)
Count monthly	23.7764	0.5931	−34.7871
	(30.1053)	(0.5154)	(108.8222)
Farmers union patrols	13.1396	0.1899	14.5626
	(17.6982)	(0.2662)	(66.8547)
Neighbourhood watch patrols	-8.0858	0.2093	-116.6720*

Table 2. *Cont.*

Variable	Tobit	Probit	Truncated
	(17.5835)	(0.2781)	(71.2700)
Private company patrols	20.5376	0.1392	48.3712
	(19.6144)	(0.2961)	(68.7394)
No Patrols	14.4166	0.1087	12.5930
	(22.0436)	(0.3254)	(85.0541)
Goodness of Fit			
No. of observations	292	292	292
Sigma	81.1383****	—	106.2761****
	(4.6923)	—	(12.4284)
Log likelihood	-1046.2581	-163.0667	-815.9966
% Correct predictions	—	69.63%	—
McFadden R^{2a}	—	0.1855	—
Model chi-square [b]	—	74.2947	—
Significance level [c]	—	(0.0000)	—
Likelihood-ratio test for Tobit vs. Truncated regression[d]	—	—	134.3896[d]
			(0.0000)[c]

****, **, * denote 1%, 5% and 15% significance level respectively. Standard errors are in parentheses. [a] McFadden R^2 is given by one minus the ratio of the unrestricted, to restricted log likelihood function values. [b] The chi-square test evaluates the null hypothesis that all coefficients (not including the constant) are jointly zero. [c] Numbers in parentheses are associated with chi-square probabilities. [d] The likelihood ratio test is given by $\lambda = 2 (\ln L_{Probit} + \ln L_{Truncated\ regression} - \ln L_{Tobit})$.

Similar to the findings for the external factors (Table 1), results from the internal factors (Table 2) indicate that the log-likelihood test ratio of 134.39 is highly significant ($p < 0.01$). Therefore, the Tobit specifications are relaxed in favour of the more general Graggs model. Thus, internal factors affecting the occurrence of livestock theft are significantly different from the factors affecting the level of livestock theft. If the Tobit model were to be used, it would fail to identify the correct factors affecting livestock theft.

Eight of the internal factors have a significant relationship with the occurrence (Probit) of livestock theft in the Free State Province. The use of livestock "guards" ($p < 0.15$) proved to be the only significant management variable positively related to the occurrence of stock theft. Thus, farmers who have a higher probability of experiencing livestock theft are making use of guards. Thus, farmers who experienced livestock theft on a regular basis have started to use guards in an attempt to control livestock theft. The only significant physical barrier variable affecting the occurrence of livestock theft is "corralling at night" ($p < 0.01$). The positive sign of the coefficient would imply that farmers who have a higher probability of experiencing livestock theft are corralling at night. This could be similar to the use of guards where the sheep are corralled in an attempt to control livestock theft. Three of the technology factors proved to be positively related to the occurrence of livestock theft and only one negatively. "stock theft collars" $p < 0.01$), "alarms in corral" ($p < 0.05$) and "strategic stock theft collars" ($p < 0.05$) were positively related, while "light in corral" ($p < 0.15$) was negatively related to the occurrence of livestock theft. The results suggest that farmers who are more likely to experience livestock theft used stock theft collars. It does not matter whether the stock theft collars are used actively or strategically. Farmers who are more likely to experience livestock theft placed alarms in their corrals and farmers who have light in their corrals are less likely to experience livestock theft. It seems that farmers are using stock theft collars and alarms because of regular losses to stock theft and that where lights are placed in corrals, it has led to lower occurrence rates of livestock theft.

None of the animals used to control livestock theft proved to have a significant relationship with the occurrence of livestock theft. Although it was hypothesised that many of the actions taken by farmers could influence the occurrence of livestock theft, only two proved to be significant. "counting animals on a daily basis" ($p < 0.15$) was negatively related to the occurrence of livestock theft and "counting animals more than once a week" ($p < 0.05$) had a positive relationship to the occurrence of livestock theft. Thus, farmers who count their animals on a daily basis are less likely to experience livestock theft and farmers who count two to three times a week are more likely to experience livestock theft. The results suggest that farmers who count on a regular basis have a lower probability for the occurrence of livestock theft.

Results from the Truncated regression show (Table 2) that nine of the internal factors have a significant relationship with the level of livestock theft experienced by farmers in the Free State Province. Management practices that have a significant relationship with the level of stock theft experienced by farmers are "theft informant" ($p < 0.01$) and "strategic theft informant" ($p < 0.10$). Taking into account the positive sign of the coefficient, farmers who are more likely to experience a higher level of livestock theft make use of a stock theft informant (both actively and strategically). "strategic collars" ($p < 0.01$) is the only physical barrier significantly related to the level of livestock theft experienced. The positive sign shows those farmers who have a probability of experiencing a higher level of livestock theft corral their animals during strategic times of the year. Two of the technologies used to control livestock theft were significant. Both "stock theft collars" ($p < 0.05$) and "cameras" ($p < 0.01$) proved to have a positive relationship to the level of livestock theft experienced. Thus, farmers who are more likely to experience a higher level of livestock theft use stock theft collars and farmers who have a higher probability of experiencing a higher level of livestock theft use cameras in and around their corrals.

As in the case of the occurrence of livestock theft (Probit), none of the animals used to control livestock theft proved to have a significant effect on the level of livestock theft experienced. Results show that the actions taken against stock theft contain four significant factors. "Active patrols" ($p < 0.01$) and "strategic access control" ($p < 0.10$) had a positive coefficient, implying that farmers who experience a higher level of livestock theft patrols throughout the year and farmers who are more likely to experience higher levels of stock theft control access to their farms during known troublesome times. "access control" ($p < 0.15$) and "neighbourhood watch patrols" ($p < 0.10$) have had negative relationships with the level of livestock theft experienced. This implies that farmers who have access control to their farms and farmers who take part in neighbourhood watches experience lower levels of livestock theft.

From a management point of view, it seems that farmers who count more often have a lower probability of experiencing livestock theft than those who count less often. This could be due to the fact that a farmer who counts his animals more often will become aware of theft at an earlier stage and thieves will have less time to get rid of the animals and/or evidence in their possession. Stock theft collars proved to be significantly related to the occurrence as well as the level of livestock theft with positive coefficients in both cases. Thus, farmers who have higher probability for the occurrence as well as the level of livestock theft, use stock theft collars. This could be an indication of how desperate the farmers who lose large numbers of livestock on a regular basis are to find a control method that works.

The signs and coefficients of the regression analyses shown in Tables 1 and 2 suggest that farmers should count their animals on a daily basis to become aware of thefts as soon as possible.

4. Conclusions

Investigation of the external factors proved that eight external factors were associated with the occurrence of livestock theft and six external factors showed a significant relationship to the level of livestock theft experienced. Results show that farmers who report their incidents of crime in any of the offered time slots increase their probability of experiencing livestock theft. However, farmers who have a higher probability of experiencing stock theft and a higher level of stock theft report their cases 13.00–24.00 h after the animals are stolen. The results thus suggest that farmers who took longer to report their cases were more likely to experience stock theft and farmers who experienced stock theft at higher level on a regular basis took longer to report. It could be that those farmers who lost large numbers of animals on a regular basis are fed up with the thefts and probably feel that it would not help to report the cases as early as possible. Interesting to note is that plains proved to be significantly related to a higher occurrence rate of livestock theft and negatively related to the level of livestock theft experienced. The results suggest that it is easier to steal one or two sheep in a flat environment; however, it is hard to conceal a large number of sheep at a time. Thus, thefts occur on a regular basis in small quantities on flatter land, whereas more mountainous areas create the opportunity to steal a larger number of animals on a less frequent basis. Strangely, it seems that farms bordering Lesotho experience stock theft on a more regular basis, but not necessarily on a larger scale than the rest of the Free State Province.

When focussing on the internal factors, eight internal factors had a significant relationship to the occurrence of livestock theft, while nine internal factors were associated with the level of livestock theft experienced. Moreover, these results showed that factors (external and internal) affecting the occurrence of livestock theft and factors affecting the level of livestock theft are different. Thus, the results from this study relate to the results that van Niekerk [17] and Badenhorst [18] reported for predation management.

Acknowledgments: The authors would like to acknowledge the Red Meat Research and Development Trust (RMRDT) who provided the funding for the collection of the data.

Author Contributions: W.A. Lombard and Antonie Geyer conceived and designed the experiments; W.A. Lombard and Walter van Niekerk performed the experiments; W.A. Lombard and Henry Jordaan analysed the data; Henry Jordaan contributed reagents/materials/ analysis tools; W.A. Lombard, Walter van Niekerk and Henry Jordaan wrote the paper. All authors have read and approved the final manuscript.

Conflicts of Interest: The authors declare no conflict of interest.

References

1. Parliamentary Monetary Group (PMG). An overview of stock theft in South Africa. Meeting report from 25 May 2010. PMG. Available online: http://www.pmg.org.za (accessed on 10 October 2013).
2. Clack, W.J. The extent of stock theft in South Africa. *Acta Criminol.: South. Afr. J. Criminol.* **2013**, *26*, 77–91.
3. Alberti, L. *Account of the Xhosa(1811)*; Fehr, W., Ed.; Balkema: Cape Town, South Africa, 1968.
4. Peires, J. Unsocial bandits: The stock thieves of Qumbu and their enemies. Democracy popular precedents practice culture. 13–15 July 1994, University of Witwatersrand, History workshop, 1994. Available online: http://wiredspace.wits.ac.za/handle/10539/ 8020 (accessed on 11 October 2013).
5. Khoabane, S.; Black, P. On the economic effects of livestock theft in Lesotho: An asset-based approach. *J. Dev. Agric. Econ.* **2012**, *4*, 142–146.
6. Anderson, D. Stock theft and moral economy in colonial Kenya. *Africa: J. Int. Afr. Inst.* **1986**, *56*, 399–416. [CrossRef]
7. Cheserek, G.J.; Omondi, P.; Odenyo, V.A.O. Nature and Causes of Cattle Rustling among some Pastoral Communities in Kenya. *J. Emerg. Trends Manag. Sci.* **2012**, *3*, 173–179.
8. Bunei, E.K.; Rono, J.K.; Chessa, S.R. Factors influencing farm crime in Kenya: Opinions and experiences of farmers. *Int. J. Rural Criminol.* **2013**, *2*, 75–100.

9. Mohammed, M.A.; Ortmann, G.F. Factors influencing adoption of livestock insurance by commercial dairy farmers in three Zobatat of Eritrea. *Agrekon* **2005**, *44*, 172–186. [CrossRef]

10. Olowa, W.O. The Effects of Livestock Pilferage on Household Poverty in Developing Countries: Theoretical Evidence from Nigeria. *Bangladesh e-J. Sociol.* **2010**, *7*, 42–46.

11. Anderson, K.M.; McCall, M. *Farm Crime in Australia*; Australian Institute of Criminology: Canberra, Australia, 2005.

12. Barclay, E.; Donnermeyer, J.F. Crime in regional Australia. In Proceedings of the 4th National Outlook Symposium on Crime in Australia, New Crimes or New Responses Convened by the Australian Institute of Criminology, Canberra, 21–22 June 2001; Australian Institute of Criminology: Canberra, Australia, 2001.

13. Dzimba, J.; Matooane, M. The impact of stock theft on human security. In *Strategies for Combating Stock Theft in Lesotho*; Kariri, J.N., Mistry, D., Eds.; Institute for Security Studies: Pretoria, South Africa, 2005.

14. National Stock Theft Prevention Forum (NSTPF; Pretoria, South Africa). South African livestock theft numbers 2010/11–2013/14. Unpublished data. 2014.

15. Red Meat Producers Organization (RPO). Business Plan and Budget for Production Development: 2012–2014, 2012. Red Meats South Africa. Available online: http://www.redmeatsa.co.za/wp-content/uploads/2014/09/RPO-Budget-and-Business-Plan-28Levy-Period-2012-201429-Production-Development9.pdf (accessed on 23 February 2015).

16. Red Meat Producers Organization (RPO). Diere van R750 Miljoen jaarliks gesteel. *Landbou Weekblad*, 26 September 2014.

17. Van Niekerk, H.N. The Cost of Predation on Small Livestock in South Africa by Medium-Sized Predators. Master's Thesis, University of the Free State, Bloemfontein, South Africa, November 2010.

18. Badenhorst, C.G. The Economic Cost of Large Stock Predation in The North West province of South Africa. Master's Thesis, University of the Free State, Bloemfontein, South Africa, July 2014.

19. Bureau for Food and Agricultural Policy (BFAP). Baseline Agricultural Outlook 2014–2023. BFAP. Available online: http://www.bfap.co.za/images/documents/baseline/bfap_baseline_2014.pdf (accessed on 20 February 2015).

20. Lesotho. The official website of Lesotho, 2015. Available online: www.gov.ls (accessed on 6 January 2015).

21. Steinberg, J. *The Lesotho/Free State Border*; Institute for Security Studies: Pretoria, South Africa, 2005; ISS Paper 113.

22. Statistics South Africa. Census 2011 Municipal factsheet, 2011. Statistics South Africa. Available online: http://www.statssa.gov.za/census/census_2011/census_products/Census_2011_Municipal_fact_sheet.pdf (accessed on 22 December 2016).

23. Statistics South Africa. Gross domestic product. Statistical release, 2014. Statistics South Africa. Available online: http://www.statssa.gov.za/publications/P0441/P04413rdQuarter2014.pdf (accessed on 22 December 2016).

24. Department of Agriculture, Forestry and Fisheries (DAFF) South Africa. Abstract of Agricultural Statistics 2013, 2013. DAFF. Available online: http://www.nda.agric.za/docs/statsinfo/Abstact2013.pdf (accessed on 22 December 2016).

25. Department of Agriculture, Forestry and Fisheries (DAFF; Pretoria, South Africa). Livestock statistics per magisterial districts in the Free State. Unpublished data. 2014.

26. Google Maps. Map for the Free State Province, 2015. Available online: https://www.google.co.za/maps/place/Free+State/@-28.6629083,24.8220063,7z/data=!3m1!4b1!4m5!3m4!1s0x1e8fc56133d61419:0xedf58a554da20a47!8m2!3d-28.4541105!4d26.7967849 (accessed on 15 December 2015).

27. Moberly, R.L. The cost of fox predation to agriculture in Britain. Ph.D. Thesis, Environment Department, University of York, York, UK, March 2002.

28. Cochran, W.G. *Sampling Techniques*, 3rd ed.; John Wiley & Sons: New York, NY, USA, 1977.

29. Fienberg, S.E. Notes on Stratified Sampling for Statistics 36–303: Sampling, Surveys, and Society. In Presented at the Department of Statistics, Carnegie Mellon University, Pittsburgh, PA, USA, 12 March 2003; Available online: http://www.stat.cmu.edu/~fienberg/Stat36-303-03/Handouts/StratificationNotes-03.pdf (accessed on 22 December 2016).

30. Lin, T.; Schmidt, P. A test of the Tobit specification against an alternative suggested by Cragg. *Rev. Econ. Stat.* **1984**, *66*, 174–177. [CrossRef]

31. Cragg, J.G. Some statistical models for limited dependent variables with application to the demand for durable goods. *Econometrica* **1971**, *39*, 829–844. [CrossRef]

32. Katchova, A.L.; Miranda, M.J. *Two-Step Econometric Estimation of Farm Characteristics Affecting Marketing Contract Decisions*; American Agricultural Economics Association: Milwaukee, WI, USA, 2004.

33. Greene, J.S.; Woodruff, R.A.; Tueller, T.T. Livestock-guarding dogs for predator control: Costs, benefits, and practicality. *Wildl. Soc. Bull.* **1984**, *12*, 44–50.

MDPI AG
St. Alban-Anlage 66
4052 Basel, Switzerland
Tel. +41 61 683 77 34
Fax +41 61 302 89 18
http://www.mdpi.com

www.ingramcontent.com/pod-product-compliance
Lightning Source LLC
Chambersburg PA
CBHW051559190326
41458CB00029B/6479